AF522370

IIlustrated Special Veterinary Pathology

IIlustrated Special Veterinary Pathology

Balaji Yadav Maddina

RANDOM PUBLICATIONS
NEW DELHI (INDIA)

IIlustrated Special Veterinary Pathology

ISBN 978-93-5111-885-5

Published in 2016 in India by

Reprinted 2025

RANDOM PUBLICATIONS

4376-A/4B, Gali Murari Lal, Ansari Road
New Delhi-110 002
Phone : +9111-43580356, 011-23289044, 011-43142548
e-mail: sales@randompublications.com,
info@randompublications.com, randomexports@gmail.com

Type Setting by : Friends Media, Delhi-110089

Digitally Printed at: Replika Press Pvt. Ltd.

Preface

The field veterinarians will also find it useful in their day to day routine work of diagnosing the diseases, identification of the causative agents and thereby suggesting the effective therapeutic and preventive measure to contain the livestock and poultry diseases.

Veterinary pathology is a specialty within veterinary medicine focusing on the diagnosis of disease. Veterinary pathologists may perform biopsies, test animal tissue and consult with veterinarians on diseases.

The control of animal diseases and the promotion and protection of animal health are essential components of any effective animal breeding and production programme. Despite remarkable technical advances in the diagnosis, prevention and control of animal diseases, the condition of animal health throughout the developing world remains generally poor, causing substantial economic losses and hindering any improvement in livestock productivity.

Adaptive immunity (also called acquired, or specific, immunity) consists of mechanisms that are stimulated by (adapt to) microbes and are capable of also recognising nonmicrobial substances, called antigens. Innate immunity is the first line of defence, because it is always ready to prevent and eradicate infections. Adaptive immunity develops later after exposure to microbes and is even more powerful in combating infections.

Thus, the book will be useful to the students as textbook of special pathology courses and to the veterinarians for better understanding the diseases of animals and birds.

– Author

Contents

1

Diseases of Immunity

INTRODUCTION

Although vital to survival, the immune system is similar to the proverbial two-edged sword. On the one hand, immunodeficiency states render humans easy prey to infections and possibly tumors; on the other hand, a hyperactive immune system may cause fatal disease, as in the case of an overwhelming allergic reaction to the sting of a bee. In yet another series of derangements, the immune system may lose its normal capacity to distinguish self from non-self, resulting in immune reactions against one's own tissues and cells (*autoimmunity*). This chapter considers diseases caused by too little immunity as well as those resulting from too much immunologic reactivity. We also consider amyloidosis, a disease in which an abnormal protein, derived in some cases from fragments of immunoglobulins, is deposited in tissues. First, we review some advances in the understanding of innate and adaptive immunity and lymphocyte biology, then give a brief description of the histocompatibility genes because their products are relevant to several immunologically mediated diseases and to the rejection of transplants.

INNATE AND ADAPTIVE IMMUNITY

The physiologic function of the immune system is to protect individuals from infectious pathogens. The mechanisms that are responsible for this protection fall into two broad categories. *Innate immunity* (also called natural, or native, immunity) refers to defence mechanisms that are present even before infection and have evolved to specifically recognise microbes and protect multicellular organisms against infections. *Adaptive immunity* (also called acquired, or specific, immunity) consists of mechanisms that are stimulated by (adapt to) microbes and are capable of also recognising nonmicrobial substances, called *antigens*. Innate immunity is the first line of defence, because it is always ready to prevent and eradicate infections. Adaptive immunity develops later after exposure to microbes and is even more powerful in combating infections. By convention, the term "immune response" refers to adaptive immunity.

The major components of innate immunity are epithelial barriers that block entry of environmental microbes, phagocytic cells (mainly neutrophils and macrophages), natural killer (NK) cells, and several plasma proteins, including the proteins of the complement system. Phagocytes are recruited to sites of infection, resulting in inflammation, and here the cells ingest the microbes and are then activated to destroy the ingested pathogens. Phagocytes recognise microbes by several membrane receptors.

These include receptors for mannose residues and N-formyl methionine-containing peptides, which are produced by microbes but not by host cells, and a family of receptors that are homologous to a *Drosophila* protein called Toll. Different Toll-like receptors (TLRs) are involved in responses to different microbial products.

Upon recognition of the relevant microbial structure, the TLRs signal by a common pathway that leads to the activation of transcription factors, notably NF-κB (nuclear factor κB). NF-êB stimulates production of cytokines and several proteins that are responsible for the microbicidal activities of the phagocytes. Phagocytes internalise microbes into vesicles, where the microbes are destroyed by reactive oxygen and nitrogen intermediates and hydrolytic enzymes.

Complement proteins, are some of the most important plasma proteins of the innate immune system. Recall that in innate immunity, the complement system is activated by binding to microbes using the alternative and lectin pathways; in adaptive immunity, it is activated by binding to antibodies using the classical pathway.

Mammalian cells express regulatory proteins that prevent inappropriate complement activation. Other circulating proteins of innate immunity are mannose-binding lectin and C-reactive protein, both of which coat microbes for phagocytosis and complement activation. Lung surfactant is also a component of innate immunity, providing protection against inhaled microbes.

The adaptive immune system consists of lymphocytes and their products, including antibodies.

The receptors of lymphocytes are much more diverse than those of the innate immune system, but lymphocytes are not inherently specific for microbes, and they are capable of recognising a vast array of foreign substances.

MESSENGER MOLECULES OF THE IMMUNE SYSTEM

The induction and regulation of immune responses involve multiple interactions among lymphocytes, monocytes, inflammatory cells (e.g., neutrophils), and endothelial cells. Many such interactions depend on cell-to-cell contact; however, many interactions and effector functions are mediated by short-acting soluble mediators, called *cytokines*. This term includes the previously designated lymphokines (lymphocyte-derived), monokines (monocyte-derived), and several other polypeptides that regulate immunologic, inflammatory, and reparative host responses. Molecularly defined cytokines are called *interleukins*, implying that they mediate communications between leukocytes. Most cytokines have a wide spectrum of effects, and some are produced by several different cell types.

A large number of cytokines have been identified by molecular cloning, and the list continues to grow. The main cytokines whose functions are well established. It is convenient to classify these mediators into distinct functional classes, although many belong to multiple categories.

- Cytokines that mediate innate (natural) immunity. Included in this group are IL-1, TNF (tumor necrosis factor, also called TNF-α), type 1 interferons, and IL-6. Some cytokines, such as IL-12 and IFN-α, are involved in both innate and adaptive immunity against intracellular microbes. Certain of these cytokines (e.g., the interferons) protect against viral infections, whereas others (e.g., IL-1 and TNF) promote leukocyte recruitment and acute inflammatory responses.
- Cytokines that regulate lymphocyte growth, activation, and differentiation. Within this category are IL-2, IL-4, IL-12, IL-15, and transforming growth factor-β (TGF-β). IL-2 is an important growth factor for T-cells, IL-4 stimulates differentiation to the T_H 2 pathway and acts on B cells as well, IL-12 stimulates differentiation to the T_H 1 pathway, and IL-15 stimulates the growth and activity of NK cells. Other cytokines in this group, such as IL-10 and TGF-β, down-regulate immune responses.
- Cytokines that activate inflammatory cells. In this category are IFN-α, which activates macrophages; IL-5, which activates eosinophils; and TNF and lymphotoxin (also called TNF-β), which induce acute inflammation by acting on neutrophils and endothelial cells.
- Cytokines that affect leukocyte movement are also called *chemokines*. Most fall into two structurally distinct subfamilies, referred to as C-C and C-X-C chemokines, on the basis of the position of cysteine (c) residues. The C-X-C chemokines are produced mainly by activated macrophages and tissue cells (e.g., endothelium), whereas the C-C

chemokines are produced largely by T cells. Different chemokines recruit different types of leukocytes to sites of inflammation. Chemokines are also normally produced in tissues and are responsible for the anatomic localisation of different cell types, for example, the location of T and B cells in distinct regions of lymphoid organs.

- Cytokines that stimulate hematopoiesis. Many cytokines derived from lymphocytes or stromal cells stimulate the growth and production of new blood cells by acting on hematopoietic progenitor cells. Several members of this family are called *colony-stimulating factors* (CSFs) because they were initially detected by their ability to promote the in vitro growth of hematopoietic cell colonies from the bone marrow. Some members of this group (e.g., GM-CSF and G-CSF) act on committed progenitor cells, whereas others, exemplified by stem cell factor (c-kit ligand), act on pluripotent stem cells.

GENERAL PROPERTIES OF CYTOKINES

Although cytokines have many diverse actions, all of them share some important properties.

- Many individual cytokines are produced by several different cell types. For example, IL-1 can be produced by virtually any cell leukocytes, endothelial cells, and fibroblasts.
- The actions of cytokines are pleiotropic, meaning that any one cytokine may act on many cell types and mediate many effects. For example, IL-2, initially discovered as a T-cell growth factor, is known to affect the growth and differentiation of B cells and NK cells as well. Cytokines are also often redundant, meaning that different cytokines may stimulate the same or overlapping biologic responses.
- Cytokines induce their effects in three ways: (1) They act on the same cell that produces them (*autocrine* effect), such as occurs when IL-2 produced by antigen-stimulated T cells stimulates the growth of the same cells; (2) they affect other cells in their vicinity (*paracrine* effect), as occurs when IL-7 produced by bone marrow or thymic stromal cells promotes the maturation of B-cell progenitors in the marrow or T-cell precursors in the thymus, respectively; and (3) they affect many cells systemically (*endocrine* effect), the best examples in this category being IL-1 and TNF, which produce the systemic acute-phase response during inflammation.
- Cytokines mediate their effects by binding to specific high-affinity receptors on their target cells. For example, IL-2 activates T cells by binding to high-affinity IL-2 receptors (IL-2R). Blockade of the IL-2R by specific antireceptor monoclonal antibodies prevents T-cell

activation. This observation is the basis for the use of anti-IL-2R antibodies to control undesirable T-cell activation, as in transplant rejection.

The knowledge gained about cytokines has practical therapeutic ramifications.

First, by inhibiting cytokine production or action, it may be possible to control the harmful effects of inflammation or tissue-damaging immune reactions. Patients with rheumatoid arthritis often show dramatic responses to TNF antagonists, an elegant example of such therapy. Second, recombinant cytokines can be administered to enhance immunity against cancer or microbial infections (immunotherapy).

DISORDERS OF THE IMMUNE SYSTEM

Having reviewed some fundamentals of basic immunology, we can now turn to general features of immunologic tissue injury and immunopathology, and some specific immunologic diseases. Our discussion is divided into four broad headings:

- *Hypersensitivity reactions*, which give rise to immunologic injury in a variety of diseases, discussed throughout this book
- *Autoimmune diseases*, which are caused by immune reactions against self
- *Immunologic deficiency syndromes*, which result from genetically determined or acquired defects in some components of the normal immune system
- *Amyloidosis*, a poorly understood disorder having immunologic association.

CELLS AND TISSUES OF THE IMMUNE SYSTEM

There are two main types of adaptive immunity—cell-mediated (or cellular) immunity, which is responsible for defence against intracellular microbes, and humoral immunity, which protects against extracellular microbes and their toxins. Cellular immunity is mediated by T (thymus-derived) lymphocytes, and humoral immunity is mediated by B (bone marrow-derived) lymphocytes and their secreted products, antibodies. All these mechanisms of adaptive immunity are capable of causing injury to the host and subsequent disease.\

T LYMPHOCYTES

T lymphocytes are generated from immature precursors in the thymus. Mature, naive T cells are found in the blood, where they constitute 60% to

70% of lymphocytes, and in T-cell zones of peripheral lymphoid organs, such as the paracortical areas of lymph nodes and periarteriolar sheaths of the spleen. The segregation of naive T cells to these anatomic sites is because the cells express receptors for chemoattractant cytokines (chemokines) that are produced only in these regions of lymphoid organs. Each T cell is genetically programmed to recognise a specific cell-bound antigen by means of an antigen-specific T-cell receptor (TCR). In approximately 95% of T cells, the TCR consists of a disulfide-linked heterodimer made up of an α and a β polypeptide chain, each having a variable (antigen-binding) and a constant region.

The αβ TCR recognises peptide antigens that are displayed by major histocompatibility complex (MHC) molecules on the surfaces of antigen-pressenting cells. (The function of the MHC is described later.) T cells (in contrast to B cells) cannot be activated by soluble antigens; therefore, presentation of processed, membrane-bound antigens by antigen-presenting cells is required for induction of cell-mediated immunity. Each TCR is noncovalently linked to a cluster of five polypeptide chains, three of which form the CD3 molecular complex and two are a dimer of the æ chain. The CD3 and æ proteins are invariant. They do not bind antigen but are involved in the transduction of signals into the T cell after the TCR has bound the antigen. T-cell receptors are capable of recognising a very large number of peptides; each T cell expresses TCR molecules of one structure and specificity. TCR diversity is generated by somatic rearrangement of the genes that encode the TCR chains.

As might be expected, every somatic cell has TCR genes from the germ line. Rearrangements of these genes occur only in T cells during their development in the thymus; hence the *presence of TCR gene rearrangements demonstrated by molecular analysis is a marker of T-lineage cells*. Such analyses are used in classification of lymphoid malignancies. Furthermore, because each T cell has a unique DNA rearrangement (and hence a unique TCR), it is possible to distinguish polyclonal (non-neoplastic) T-cell proliferations from monoclonal (neoplastic) T-cell proliferations.

A minority of mature T cells express another type of TCR composed of α and ä polypeptide chains. The γδ TCR recognises peptides, lipids, and small molecules, without a requirement for display by MHC proteins. αä T cells tend to aggregate at epithelial surfaces, such as the mucosa of the respiratory and gastrointestinal tracts, suggesting that these cells are sentinels that protect against microbes that try to enter through these epithelia. However, the precise functions of γδ T cells are not known. Another small subset of T cells expresses markers that are found on natural killer (NK) cells; these cells are called NK-T cells. NK-T cells express a very limited diversity of

TCRs, and they recognise glycolipids that are displayed by the MHC-like molecule CD1. The functions of NK-T cells are also not well defined.

In addition to CD3 and æ proteins, T cells express a number of nonpolymorphic, function-associated molecules, also called accessory molecules, including CD4, CD8, CD2, integrins, and CD28. CD4 and CD8 are expressed on two mutually exclusive subsets of αβ T cells. CD4 is expressed on approximately 60% of mature CD3+ T cells, whereas CD8 is expressed on about 30% of T cells.

These T-cell membrane-associated glycoproteins serve as coreceptors in T-cell activation. During antigen presentation, CD4 molecules bind to the nonpolymorphic portions of class II MHC molecules expressed on antigen-presenting cells. In contrast,

CD8 molecules bind to class I MHC molecules. CD4 and CD8 are required to initiate signals that activate T cells that recognise antigens. Because of this requirement for coreceptors, CD4+ helper T cells can recognise and respond to antigen only in the context of class II MHC molecules, whereas CD8+ cytotoxic T cells recognise cell-bound antigens only in association with class I MHC molecules. It is now well established that T cells need two signals for activation. Signal 1 is provided when the TCR is engaged by the appropriate MHC-bound antigen, and the coreceptors CD4 and CD8 bind to MHC molecules. Signal 2 is delivered by the interaction of the CD28 molecule on T cells with the costimulatory molecules B7-1 (CD80) and B7-2 (CD86) expressed on antigen-presenting cells.

The importance of co-stimulation by this pathway is attested to by the fact that, in the absence of signal 2, the T cells fail to respond, undergo apoptosis, or become unreactive. When T cells are activated by antigen and costimulators, they secrete locally acting proteins called *cytokines*. Under the influence of a cytokine called interleukin-2 (IL-2), the T cells proliferate, thus generating a large number of antigen-specific lymphocytes. Some of these cells differentiate into effector cells, which perform the function of eliminating the antigen that started the response. Other activated cells differentiate into memory cells, which are long-lived and poised to respond rapidly to repeat encounters with the antigen.

CD4+ and CD8+ T cells perform distinct but somewhat overlapping effector functions. The CD4+ T cell can be viewed as a master regulator—the conductor of a symphony orchestra, so to speak. By secreting cytokines, CD4+ T cells influence the function of virtually all other cells of the immune system, including other T cells, B cells, macrophages, and NK cells. The central role of CD4+ T cells is tragically illustrated when the human immunodeficiency virus cripples the immune system by selective destruction of this T-cell subset.

In recent years, two functionally distinct populations of CD4+ helper cells have been recognised on the basis of the different cytokines they produce. The T-helper-1 (T_H 1) subset synthesises and secretes IL-2 and interferon-α (IFN-α) but not IL-4 or IL-5, whereas T_H 2 cells produce IL-4, IL-5 and IL-13 but not IL-2 or IFN-γ. This distinction is significant because the cytokines secreted by these subsets have different effects on other immune cells. The T_H 1 subset is involved in facilitating delayed hypersensitivity, macrophage activation, and synthesis of opsonising and complement-fixing antibodies, such as IgG2a in mice, all of which are actions of IFN-α. The T_H 2 subset aids in the synthesis of other classes of antibodies, notably IgE (mediated by IL-4 and IL-13) and in the activation of eosinophils (mediated by IL-5). CD8+ T cells function mainly as cytotoxic cells to kill other cells but, similar to CD4+ T cells, they can secrete cytokines, primarily of the T_H 1 type.

Macrophages

Macrophages are a part of the mononuclear phagocyte system; their origin, differentiation, and role in inflammation. Here we need only to emphasise that macrophages play important roles both in the induction and in the effector phase of immune responses.

- Macrophages that have phagocytosed microbes and protein antigens process the antigens and present peptide fragments to T cells. Thus, macrophages are involved in the induction of cell-mediated immune responses.
- Macrophages are important effector cells in certain forms of cell-mediated immunity, such as the delayed hypersensitivity reaction. As mentioned earlier, macrophages are activated by cytokines, notably IFN-γ produced by the T_H 1 subset of CD4+ cells. Such activation enhances the microbicidal properties of macrophages and augments their ability to kill tumor cells.
- Macrophages are also important in the effector phase of humoral immunity. The macrophages phagocytose microbes that are opsonized (coated) by IgG or C3b.

Natural Killer Cells

NK cells make up approximately 10% to 15% of the peripheral blood lymphocytes and do not bear T-cell receptors or cell surface immunoglobulins. Morphologically, NK cells are somewhat larger than small lymphocytes, and they contain abundant azurophilic granules. Hence, they are also called *large granular lymphocytes*. NK cells are endowed with an innate ability to kill a variety of tumor cells, virally infected cells, and some normal cells, without previous sensitisation. These cells are part of the innate immune system, and they may be the first line of defence against viral infections and, perhaps, some tumors. NK cells do not rearrange T-cell receptor genes and are CD3 negative. Two cell surface molecules, CD16 and CD56, are widely used to identify NK cells. CD16 is the Fc receptor for IgG and it endows NK cells with another function, the ability to lyse IgG-coated target cells. This phenomenon, known as *antibody-dependent cell-mediated cytotoxicity*.

The functional activity of NK cells is regulated by a balance between signals from activating and inhibitory receptors. The activating receptors stimulate NK cell killing by recognising ill-defined molecules on target cells, some of which may be viral products; the inhibitory receptors inhibit the activation of NK cells by recognition of self-class I MHC molecules. The class I MHC-recognising inhibitory receptors on NK cells are aptly called killer inhibitory receptors.

They are biochemically distinct from T-cell receptors. It is believed that NK cells are inhibited from killing normal cells because all nucleated normal cells express self-class I MHC molecules. If virus infection or neoplastic transformation perturbs or reduces the expression of class I MHC molecules, inhibitory signals delivered to NK cells are interrupted, and lysis occurs. However, merely the absence of inhibition is not sufficient for NK cell-mediated killing, and NK cell-mediated killing requires triggering of activating receptors in conjunction with release of inhibitory receptors. Several types of activating receptors have been discovered, including members of the NKG2D family and some Ig-like receptors.

The NKG2D receptors recognise stress-induced proteins that are normally expressed by only a few cells in the gut epithelium but whose expression increases on many cells following viral infection or neoplastic transformation. Other activating receptors recognise viral proteins that are structurally similar to class I MHC molecules.

Thus, NK cells are activated by contact with virus-infected and tumor cells, both of which often express reduced levels of class I MHC molecules and therefore do not engage inhibitory receptors. NK cells also secrete cytokines, such as IFN-γ, TNF, and granulocyte macrophage colony-stimulating factor (GM-CSF). IFN-γ activates macrophages to destroy ingested microbes, and thus NK cells provide early defence against intracellular microbial infections. IFN-γ also promotes the differentiation of naive CD4+ T-cells into T_H 1 cells. Thus, activation of NK cells early in the immune response can favour induction of delayed hypersensitivity and secretion of opsonising antibodies by promoting the development of T_H 1 cells. The activity of NK cells is regulated by many cytokines, including IL-2, IL-15, and IL-12. IL-2 and IL-15 stimulate proliferation of NK cells, whereas IL-12 activates killing and secretion of IFN-γ.

STRUCTURE AND FUNCTION OF HISTOCOMPATIBILITY MOLECULES

Although originally identified as antigens that evoke rejection of transplanted organs, histocompatibility molecules are now known to be extremely important for the induction and regulation of the immune response. *The principal physiologic function of the cell surface histocompatibility molecules is to bind peptide fragments of foreign proteins for presentation to antigen-specific T cells.* Recall that T cells (in contrast to B cells) can recognise only membrane-bound antigens, and hence histocompatibility molecules are critical to the induction of T-cell immunity. Here we summarise the salient features of human histocompatibility molecules, primarily to facilitate understanding of their role in rejection of organ transplants and in disease susceptibility. In humans, the genes encoding the most important histocompatibility molecules are clustered on a small segment of chromosome 6, the *major histocompatibility complex*, or the *human leukocyte antigen* (HLA) complex in humans, so named because MHC-encoded antigens were initially detected on leukocytes. The HLA system is highly polymorphic, meaning that there are many alleles of each MHC gene in the population and each individual inherits one (often unique) set of these alleles. This, as we see subsequently, constitutes a formidable barrier in organ transplantation.

On the basis of their chemical structure, tissue distribution, and function, the MHC gene products are classified into three categories. Class I and class II genes encode cell surface glycoproteins involved in antigen presentation. Class III genes encode components of the complement system.

Class I MHC molecules are expressed on all nucleated cells and platelets. They are encoded by three closely linked loci, designated HLA-A, HLA-B, and HLA-C. Each of these molecules is a heterodimer, consisting of a polymorphic α, or heavy, chain (44-kD) linked noncovalently to a smaller (12-kD) nonpolymorphic peptide called *β_2-microglobulin*, which is not encoded within the MHC.

The extracellular region of the heavy chain is divided into three domains: α_1, α_2, and α_3. Crystal structure of class I molecules has revealed that the α_1 and α_2 domains form a cleft, or groove, where peptides bind to the MHC molecule. Biochemical analyses of several different class I alleles have revealed that almost all polymorphic residues line the sides or the base of the peptide-binding groove. As a result, different class I alleles bind and display different peptide fragments. In general, class I MHC molecules bind and display peptides that are derived from proteins, such as viral antigens, synthesised within the cell.

The generation of peptide fragments within the cells, and their association with MHC molecules and transport to the cell surface, is a complex process. Involved in this sequence are proteolytic complexes (proteasomes), which digest antigenic proteins in the cytoplasm into short peptides, and transport proteins, which ferry peptide fragments from the cytoplasm to the endoplasmic reticulum. Within the endoplasmic reticulum, peptides bind to the antigen-binding cleft of newly synthesised class I heavy chains, which then associate with β_2 -microglobulin to form a stable trimer that is transported to the cell surface for presentation to CD8+ cytotoxic T lymphocytes.

In this interaction, the TCR recognises the MHC-peptide complex, and the CD8 molecule, acting as a coreceptor, binds to the nonpolymorphic α_3 domain of the class I heavy chain. CD8+ cytotoxic T cells can recognise viral (or other) peptides only if presented as a complex with self-class I antigens, and therefore CD8+ T cells are said to be *class I MHC-restricted*. In the eyes of T cells, self-MHC molecules are those that they "grew up with" during maturation within the thymus. Because one of the important functions of CD8+ T cells is to eliminate viruses, which may infect any nucleated cell, it makes good sense to have widespread expression of class I HLA molecules.

Class II MHC molecules are coded for in a region called *HLA-D*, which has three subregions: HLA-DP, HLA-DQ, and HLA-DR. Each class II molecule is a heterodimer consisting of a noncovalently associated α chain and β chain. Both chains are polymorphic, and each of the three HLA-D subregions encodes one α chain and one β chain. The extracellular portions of the α and β chains have two domains each: α_1, α_2 and β_1, β_2. Crystal structure of class II molecules has revealed that, similar to class I molecules, they have an antigen-binding cleft facing outward. In contrast to class I molecules, however, the antigen-binding cleft is formed by an interaction of the α_1 and β_1 domains of both chains, and it is in this portion that most class II alleles differ.

Thus, it seems that, as with class I molecules, polymorphism of class II molecules is associated with differential binding of antigenic peptides. The nature of peptides that bind to class II molecules is different from that of peptides that bind to class I molecules. In general, class II molecules present exogenous antigens (e.g., extracellular microbes, soluble proteins) that are first internalised and processed in the endosomes or lysosomes. Peptides resulting from proteolytic cleavage then associate with class II heterodimers that were assembled in the endoplasmic reticulum and transported into the vesicles. Finally, the peptide-MHC complex is transported to the cell surface, where it can be recognised by CD4+ helper T cells. In this interaction, the CD4 molecule acts as the coreceptor.

Because CD4+ T cells can recognise antigens only in the context of self-class II molecules, they are referred to as *class II MHC-restricted*. In contrast to class I molecules, the tissue distribution of MHC class II molecules is largely restricted to antigen-presenting cells (macrophages, dendritic cells, and B cells). Expression of class II molecules can be induced on several other cell types, however, including endothelial cells and fibroblasts, by the action of IFN-α.

MHC molecules play key roles in regulating T cell-mediated immune responses in two ways. First, because different antigenic peptides bind to different class II gene products, it follows that an individual mounts a vigorous immune response against an antigen only if he or she inherits the gene(s) for those class II molecule(s) that can bind the antigen and present it to helper T cells. The consequences of inheriting a given class II gene depend on the nature of the antigen bound by the class II molecule.

For example, if the antigen is a peptide from ragweed pollen, the individual who expresses class II molecules capable of binding the antigen would be genetically prone to allergic reactions against pollen. In contrast, an inherited capacity to bind a bacterial peptide may provide resistance to disease by evoking a protective antibody response.

Second, during their maturation in the thymus, only T cells that can recognise self-MHC molecules are selected for export to the periphery. Thus, the type of MHC molecules that T cells encounter during their development influences the reactivity of mature peripheral T cells.

HLA AND DISEASE ASSOCIATION

A variety of diseases have been found to be associated with certain HLA alleles. The best known is the association between ankylosing spondylitis and HLA-B27; individuals who inherit this allele have a 90-fold greater chance (relative risk) of developing the disease than those who are negative for HLA-B27. The diseases that show association with the HLA locus can be broadly grouped into the following categories:

1. *Inflammatory diseases*, including ankylosing spondylitis and several postinfectious arthropathies, all associated with HLA-B27
2. *Inherited errors of metabolism*, such as 21-hydroxylase deficiency (HLA-BW47) and hereditary hemochromatosis (HLA-A)
3. *Autoimmune diseases*, including autoimmune endocrinopathies, associated mainly with alleles at the DR locus.

The mechanisms underlying these associations are not fully understood. In some cases (e.g., 21-hydroxylase deficiency), the linkage results from the fact that the relevant disease-associated gene, in this case the gene for 21-hydroxylase, maps within the HLA complex. Similarly, in hereditary hemochromatosis, a gene that is mutated, called *HFE*, maps within the HLA locus. *HFE* resembles MHC molecules structurally, but its function is not in the presentation of antigens to T cells but in the regulation of iron transport. In the case of immunologically mediated disorders, it seems likely that the role of HLA class II molecules in regulating immune responsiveness may be relevant.

MECHANISMS OF HYPERSENSITIVITY REACTIONS

Humans live in an environment teeming with substances capable of producing immunologic responses. Contact with antigen leads not only to induction of a protective immune response, but also to reactions that can be damaging to tissues. Exogenous antigens occur in dust, pollens, foods, drugs, microbiologic agents, chemicals, and many blood products used in clinical practice. The immune responses that may result from such exogenous antigens take a variety of forms, ranging from annoying but trivial discomforts, such as itching of the skin, to potentially fatal diseases, such as bronchial asthma. The various reactions produced are called *hypersensitivity reactions*, and tissue injury in these reactions may be caused by humoral or cell-mediated immune mechanisms.

Injurious immune reactions may be evoked not only by exogenous environmental antigens, but also by endogenous tissue antigens. Some of these immune reactions are triggered by homologous antigens that differ among individuals with different genetic backgrounds. Transfusion reactions and graft rejection are examples of immunologic disorders evoked by homologous antigens. Another category of disorders, those incited by self-, or autologous, antigens, constitutes the important group of autoimmune diseases. These diseases arise because of the emergence of immune responses against self-antigens.

Hypersensitivity diseases can be classified on the basis of the immunologic mechanism that mediates the disease. This classification is of value in distinguishing the manner in which the immune response ultimately causes tissue injury and disease, and the accompanying pathologic alterations. Prototypes of each of these immune mechanisms are presented in the subsequent sections.

- In *immediate hypersensitivity (type I hypersensitivity)*, the immune response releases vasoactive and spasmogenic substances that act on vessels and smooth muscle and pro-inflammatory cytokines that recruit inflammatory cells.

- In *antibody-mediated disorders (type II hypersensitivity)*, secreted antibodies participate directly in injury to cells by promoting their phagocytosis or lysis and injury to tissues by inducing inflammation. Antibodies may also interfere with cellular functions and cause disease without tissue injury.
- In *immune complex-mediated disorders (type III hypersensitivity)*, antibodies bind antigens and then induce inflammation directly or by activating complement. The leukocytes that are recruited (neutrophils and monocytes) produce tissue damage by release of lysosomal enzymes and generation of toxic free radicals.
- In *cell-mediated immune disorders (type IV hypersensitivity)*, sensitised T lymphocytes are the cause of the cellular and tissue injury.

Immediate (Type I) Hypersensitivity

Immediate, or type I, hypersensitivity is a rapidly developing immunologic reaction occurring within minutes after the combination of an antigen with antibody bound to mast cells in individuals previously sensitised to the antigen. These reactions are often called allergy, and the antigens that elicit them are allergens. Immediate hypersensitivity may occur as a systemic disorder or as a local reaction.

The systemic reaction usually follows injection of an antigen to which the host has become sensitised. Often within minutes, a state of shock is produced, which is sometimes fatal. The nature of local reactions varies depending on the portal of entry of the allergen and may take the form of localised cutaneous swellings (skin allergy, hives), nasal and conjunctival discharge (allergic rhinitis and conjunctivitis), hay fever, bronchial asthma, or allergic gastroenteritis (food allergy). Many local type I hypersensitivity reactions have two well-defined phases.

The *immediate, or initial, response* is characterised by vasodilation, vascular leakage, and depending on the location, smooth muscle spasm or glandular secretions. These changes usually become evident within 5 to 30 minutes after exposure to an allergen and tend to subside in 60 minutes. In many instances (e.g., allergic rhinitis and bronchial asthma), a *second, late-phase* reaction sets in 2 to 24 hours later without additional exposure to antigen and may last for several days. This late-phase reaction is characterised by infiltration of tissues with eosinophils, neutrophils, basophils, monocytes, and CD4+ T cells as well as tissue destruction, typically in the form of mucosal epithelial cell damage.

Because mast cells are central to the development of immediate hypersensitivity, we first review some of their salient characteristics and then discuss the immune mechanisms that underlie this form of hypersensitivity. *Mast cells* are bone marrow-derived cells that are widely distributed in the tissues.

They are found predominantly near blood vessels and nerves and in subepithelial sites, where local immediate hypersensitivity reactions tend to occur. Mast cells have cytoplasmic membrane-bound granules that contain a variety of biologically active mediators. In addition, mast-cell granules contain acidic proteoglycans that bind basic dyes such as toluidine blue. Because the stained granules often acquire a colour that is different from that of the native dye, they are referred to as *metachromatic* granules. Mast cells (and basophils) are activated by the cross-linking of high-affinity IgE Fc receptors; in addition, mast cells may also be triggered by several other stimuli, such as complement components C5a and C3a (anaphylatoxins), both of which act by binding to their receptors on the mast-cell membrane. Other mast-cell secretagogues include macrophage-derived cytokines, some drugs such as codeine and morphine, adenosine, mellitin (present in bee venom), and physical stimuli (e.g., heat, cold, sunlight).

Basophils are similar to mast cells in many respects, including the presence of cell-surface IgE Fc receptors as well as cytoplasmic granules. In contrast to mast cells, however, basophils are not normally present in tissues but rather circulate in the blood in extremely small numbers. (Most allergic reactions occur in tissues, and the role of basophils in these reactions is not as well established as that of mast cells.) Similar to other granulocytes, basophils can be recruited to inflammatory sites.

Most immediate hypersensitivity reactions are mediated by IgE antibodies. IgE-secreting B cells differentiate from naive (membrane IgM and IgD-expressing) B cells, and this process is dependent on the activity of CD4+ helper T cells of the T_H 2 type. Hence, T_H 2 cells are pivotal in the pathogenesis of type I hypersensitivity. The first step in the synthesis of IgE is the presentation of the antigen to naive CD4+ helper T cells by dendritic cells that capture the antigen from its site of entry. In response to antigen and other stimuli, including cytokines produced at the local site, the T cell differentiate into T_H 2 cells. The newly minted T_H 2 cells produce a cluster of cytokines upon subsequent encounter with the antigen; as we mentioned earlier, the signature cytokines of this subset are IL-4, IL-5, and IL-13. IL-4 is essential for turning on the IgE-producing B cells and for sustaining the development of T_H 2 cells. IL-5 activates eosinophils, which, as we discuss subsequently, are important effectors of type I hypersensitivity. IL-13 promotes IgE production and acts on epithelial cells to stimulate mucus secretion. In addition, T_H 2 cells and epithelial cells produce chemokines that attract more T_H 2 cells, as well as eosinophils and occasionally basophils, to the reaction site.

Mast cells and basophils express high-affinity receptors for the Fc portion of IgE, and therefore avidly bind IgE antibodies. When a mast cell, armed with cytophilic IgE antibodies, is re-exposed to the specific allergen, a series of reactions takes place, leading eventually to the release of a variety of powerful mediators responsible for the clinical expression of immediate hypersensitivity reactions.

In the first step in this sequence, antigen (allergen) binds to the IgE antibodies previously attached to the mast cells. Multivalent antigens bind to more than one IgE molecule and thus cross-link adjacent IgE antibodies and the underlying IgE Fc receptors.

The bridging of IgE molecules activates signal transduction pathways from the cytoplasmic portion of the IgE Fc receptors. These signals initiate two parallel and interdependent processes —one leading to mast cell degranulation with discharge of preformed (primary) mediators that are stored in the granules, and the other involving de novo synthesis and release of secondary mediators.

These mediators are directly responsible for the initial, sometimes explosive, symptoms of immediate hypersensitivity, and they also set into motion the events that lead to the late-phase response. In addition to inducing mediator release and production, signals from IgE Fc receptors promote the survival of mast cells and can enhance expression of the Fc receptor, providing an amplification mechanism.

Primary Mediators

Primary mediators contained within mast-cell granules can be divided into three categories:

- *Biogenic amines.* The most important vasoactive amine is histamine. Histamine causes intense smooth muscle contraction, increased vascular permeability, and increased secretion by nasal, bronchial, and gastric glands.
- *Enzymes.* These are contained in the granule matrix and include neutral proteases (chymase, tryptase) and several acid hydrolases. The enzymes cause tissue damage and lead to the generation of kinins and activated components of complement (e.g., C3a) by acting on their precursor proteins.

- *Proteoglycans.* These include heparin, a well-known anticoagulant, and chondroitin sulfate. The proteoglycans serve to package and store the other mediators in the granules.

Secondary Mediators

Secondary mediators include two classes of compounds (1) lipid mediators and (2) cytokines.

The *lipid mediators* are generated by sequential reactions in the mast-cell membranes that lead to activation of phospholipase A_2, an enzyme that acts on membrane phospholipids to yield *arachidonic acid*. This is the parent compound from which leukotrienes and prostaglandins are derived by the 5-lipoxygenase and cyclooxygenase pathways.

- *Leukotrienes.* Leukotrienes C_4 and D_4 are the most potent vasoactive and spasmogenic agents known. On a molar basis, they are several thousand times more active than histamine in increasing vascular permeability and causing bronchial smooth muscle contraction. *Leukotriene B_4* is highly chemotactic for neutrophils, eosinophils, and monocytes.
- *Prostaglandin D_2.* This is the most abundant mediator derived by the cyclooxygenase pathway in mast cells. It causes intense bronchospasm as well as increased mucus secretion.
- *Platelet-activating factor (PAF).* PAF is produced by some mast-cell populations. It causes platelet aggregation, release of histamine, bronchospasm, increased vascular permeability, and vasodilation. In addition, it has important pro-inflammatory actions. PAF is chemotactic for neutrophils and eosinophils. At high concentrations, it activates the newly recruited inflammatory cells, causing them to aggregate and degranulate. Because of its ability to recruit and activate inflammatory cells, it is considered important in the initiation of the late-phase response. Although the production of PAF is also triggered by the activation of phospholipase A_2, it is not a product of arachidonic acid metabolism.
- *Cytokines.* Mast cells are sources of many cytokines, which play an important role in the late-phase reaction of immediate hypersensitivity because of their ability to recruit and activate inflammatory cells. The cytokines include TNF, IL-1, IL-3, IL-4, IL-5, IL-6, and GM-CSF, as well as chemokines, such as macrophage inflammatory protein (MIP)-1α and MIP-1β. Mast cell-derived TNF and chemokines are important mediators of the inflammatory response seen at the site of allergic inflammation. Inflammatory cells that accumulate at the sites of type I hypersensitivity reactions are additional sources of cytokines and

of histamine-releasing factors that cause further mast-cell degranulation.

A final point that should be mentioned in this general discussion of immediate hypersensitivity is that *susceptibility to these reactions is genetically determined.* The term *atopy* refers to a predisposition to develop localised immediate hypersensitivity reactions to a variety of inhaled and ingested allergens. Atopic individuals tend to have higher serum IgE levels, and more IL-4-producing T_H 2 cells, compared with the general population. A positive family history of allergy is found in 50% of atopic individuals. The basis of familial predisposition is not clear, but studies in patients with asthma reveal linkage to several gene loci. Candidate genes have been mapped to 5q31, where genes for the cytokines IL-3, IL-4, IL-5, IL-9, IL-13, and GM-CSF are located, consistent with the idea that these cytokines are involved in the reactions. Linkage has also been noted to 6p, close to the HLA complex, suggesting that the inheritance of certain HLA alleles permits reactivity to certain allergens. Another asthma-associated locus is on chromosome 11q13, the location of the gene encoding the β chain of the high-affinity IgE receptor, but many studies have failed to establish a linkage of atopy with the FcepsilonRI β chain chain or even this chromosomal region.

To summarise, immediate (type I) hypersensitivity is a complex disorder resulting from an IgE-mediated triggering of mast cells and subsequent accumulation of inflammatory cells at sites of antigen deposition. These events are regulated in large part by the induction of T_H2-type helper T cells that promote synthesis of IgE and accumulation of inflammatory cells, particularly eosinophils. The clinical features result from release of mast-cell mediators as well as the accumulation of an eosinophilrich inflammatory exudate. With this consideration of the basic mechanisms of type I hypersensitivity, we turn to some conditions that are important examples of IgE-mediated disease.

Systemic Anaphylaxis

Systemic anaphylaxis is characterised by vascular shock, widespread edema, and difficulty in breathing. In humans, systemic anaphylaxis may occur after administration of foreign proteins (e.g., antisera), hormones, enzymes, polysaccharides, and drugs (such as the antibiotic penicillin). The severity of the disorder varies with the level of sensitisation. Extremely small doses of antigen may trigger anaphylaxis, for example, the tiny amounts used in ordinary skin testing for various forms of allergies. Within minutes after exposure, itching, hives, and skin erythema appear, followed shortly thereafter by a striking contraction of respiratory bronchioles and respiratory distress. Laryngeal edema results in hoarseness.

Vomiting, abdominal cramps, diarrhea, and laryngeal obstruction follow, and the patient may go into shock and even die within the hour. The risk of anaphylaxis must be borne in mind when certain therapeutic agents are administered.

Although patients at risk can generally be identified by a previous history of some form of allergy, the absence of such a history does not preclude the possibility of an anaphylactic reaction.

Local Immediate Hypersensitivity Reactions

Local immediate hypersensitivity, or allergic, reactions are exemplified by so-called atopic allergy. About 10% of the population suffers from allergies involving localised reactions to common environmental allergens, such as pollen, animal dander, house dust, foods, and the like. Specific diseases include urticaria, angioedema, allergic rhinitis (hay fever), and some forms of asthma, all discussed elsewhere in this book. The familial predisposition to the development of this type of allergy has been mentioned earlier.

Antibody-Mediated (Type II) Hypersensitivity

Type II hypersensitivity is mediated by antibodies directed towards antigens present on cell surfaces or extracellular matrix. The antigenic determinants may be intrinsic to the cell membrane or matrix, or they may take the form of an exogenous antigen, such as a drug metabolite, that is adsorbed on a cell surface or matrix. In either case, the hypersensitivity reaction results from the binding of antibodies to normal or altered cell-surface antigens. Three different antibody-dependent mechanisms involved in this type of reaction. Most of these reactions involve the effector mechanisms that are used by antibodies, namely the complement system and phagocytes.

Opsonisation and Complement- and Fc Receptor-Mediated Phagocytosis

The depletion of cells targeted by antibodies is, to a large extent, because the cells are coated (opsonised) with molecules that make them attractive for phagocytes.

When antibodies are deposited on the surfaces of cells, they may activate the complement system (if the antibodies are of the IgM or IgG class). Complement activation generates byproducts, mainly C3b and C4b, which are deposited on the surfaces of the cells and recognised by phagocytes that express receptors for these proteins. In addition, cells opsonised by IgG antibodies are recognised by phagocyte Fc receptors, which are specific for the Fc portions of some IgG subclasses. The net result is the phagocytosis of the opsonised cells and their destruction. Complement activation on cells also leads to the formation of the membrane attack complex, which disrupts membrane integrity by "drilling holes" through the lipid bilayer, thereby causing osmotic lysis of the cells.

Antibody-mediated destruction of cells may occur by another process called *antibody-dependent cellular cytotoxicity (ADCC)*. This form of antibody-mediated cell injury does not involve fixation of complement but instead requires the cooperation of leukocytes. Cells that are coated with low concentrations of IgG antibody are killed by a variety of effector cells, which bind to the target by their receptors for the Fc fragment of IgG, and cell lysis proceeds without phagocytosis. ADCC may be mediated by monocytes, neutrophils, eosinophils, and NK cells. Although, in most instances, IgG antibodies are involved in ADCC, in certain cases (e.g., eosinophil-mediated cytotoxicity against parasites), IgE antibodies are used. The role of ADCC in hypersensitivity diseases is uncertain.

Clinically, antibody-mediated cell destruction and phagocytosis occur in the following situations: (1) *transfusion reactions*, in which cells from an incompatible donor react with and are opsonised by preformed antibody in the host; (2) *erythroblastosis fetalis*, in which there is an antigenic difference between the mother and the fetus, and antibodies (of the IgG class) from the mother cross the placenta and cause destruction of fetal red cells; (3) *autoimmune hemolytic anemia*, *agranulocytosis*, and *thrombocytopenia*, in which individuals produce antibodies to their own blood cells, which are then destroyed; and (4) *certain drug reactions*, in which antibodies are produced that react with the drug, which may be attached to the surface of erythrocytes or other cells.

COMPLEMENT- AND FC RECEPTOR-MEDIATED INFLAMMATION

When antibodies deposit in extracellular tissues, such as basement membranes and matrix, the resultant injury is because of inflammation and not because of phagocytosis or lysis of cells. The deposited antibodies activate complement, generating byproducts, such as C5a (and to a lesser extent C4a and C3a), that recruit neutrophils and monocytes. The same cells also bind to the deposited antibodies via their Fc receptors. The leukocytes are activated, they release injurious substances, such as enzymes and reactive oxygen intermediates, and the result is damage to the tissues. It was once thought that complement was the major mediator of antibody-induced inflammation, but knockout mice lacking Fc receptors also show striking reduction in these reactions. It is now believed that inflammation in antibody-mediated (and immune complex-mediated) diseases is because of both complement and Fc receptor-dependent reactions.

Antibody-mediated inflammation is the mechanism responsible for tissue injury in some forms of *glomerulonephritis*, *vascular rejection* in organ grafts, and other diseases. The same reaction is involved in immune complex-mediated diseases.

Antibody-Mediated Cellular Dysfunction

In some cases, antibodies directed against cell-surface receptors impair or dysregulate function without causing cell injury or inflammation. For example, in *myasthenia gravis*, antibodies reactive with acetylcholine receptors in the motor end-plates of skeletal muscles impair neuromuscular transmission and therefore cause muscle weakness. In *pemphigus vulgaris*, antibodies against desmosomes disrupt intercellular junctions in epidermis, leading to the formation of skin vesicles.

The converse (i.e., antibody-mediated stimulation of cell function) is noted in *Graves disease*. In this disorder, antibodies against the thyroid-stimulating hormone receptor on thyroid epithelial cells stimulate the cells, resulting in hyperthyroidism.

IMMUNE COMPLEX-MEDIATED (TYPE III) HYPERSENSITIVITY

Antigen-antibody complexes produce tissue damage mainly by eliciting inflammation at the sites of deposition. The toxic reaction is initiated when antigen combines with antibody within the circulation (circulating immune complexes) and these are deposited, typically in vessel walls, or the complexes are formed at extravascular sites where antigen may have been deposited previously (in situ immune complexes).

The pathogenesis of systemic immune complex disease can be divided into three phases: (1) formation of antigen-antibody complexes in the circulation; (2) deposition of the immune complexes in various tissues, thus initiating; and (3) an inflammatory reaction at the sites of immune complex deposition. The *first phase* is initiated by the introduction of antigen, usually a protein, and its interaction with immunocompetent cells, resulting in the formation of antibodies approximately a week after the injection of the protein.

These antibodies are secreted into the blood, where they react with the antigen still present in the circulation to form antigen-antibody complexes. In the *second phase*, the circulating antigen-antibody complexes are deposited in various tissues.

The factors that determine whether immune complex formation will lead to tissue deposition and disease are not fully understood, but two possible influences are the size of the immune complexes and the functional status of the mononuclear phagocyte system:

- Large complexes formed in great antibody excess are rapidly removed from the circulation by the mononuclear phagocyte system and are therefore relatively harmless. The most pathogenic complexes are of small or intermediate size (formed in slight antigen excess), which bind less avidly to phagocytic cells and therefore circulate longer.
- Because the mononuclear phagocyte system normally filters out the circulating immune complexes, its overload or intrinsic dysfunction increases the probability of persistence of immune complexes in circulation and tissue deposition.

In addition, several other factors, such as charge of the immune complexes (anionic versus cationic), valency of the antigen, avidity of the antibody, affinity of the antigen to various tissue components, three-dimensional (lattice) structure of the complexes, and hemodynamic factors, influence the tissue deposition of complexes. Because most of these influences have been investigated with reference to deposition of immune complexes in the glomeruli. In addition to the renal glomeruli, the favored sites of immune complex deposition are joints, skin, heart, serosal surfaces, and small blood vessels. For complexes to leave the microcirculation and deposit in the vessel wall, an increase in vascular permeability must occur.

This is believed to occur when immune complexes bind to inflammatory cells through their Fc or C3b receptors and trigger release of vasoactive mediators as well as permeability-enhancing cytokines. Mast cells may also be involved in this phase of the reaction.

Once complexes are deposited in the tissues, they initiate an acute inflammatory reaction *(third phase)*. During this phase (approximately 10 days after antigen administration), clinical features such as fever, urticaria, arthralgias, lymph node enlargement, and proteinuria appear.

Wherever complexes deposit, the tissue damage is similar. Two mechanisms are believed to cause *inflammation* at the sites of deposition : (1) activation of the complement cascade, and (2) activation of neutrophils and macrophages through their Fc receptors. The *complement activation* promotes inflammation mainly by production of chemotactic factors, which direct the migration of polymorphonuclear leukocytes and monocytes (mainly C5a) and by release of anaphylatoxins (C3a and C5a), which increase vascular permeability.

Tissue damage is also mediated by oxygen free radicals produced by activated neutrophils. Immune complexes have several other effects, including aggregation of platelets and activation of Hageman factor; both of these reactions augment the inflammatory process and initiate the formation of microthrombi. The resultant inflammatory lesion is termed *vasculitis* if it occurs in blood vessels, *glomerulonephritis* if it occurs in renal glomeruli, *arthritis* if it occurs in the joints, and so on.

It is clear from the foregoing that complement-fixing antibodies (i.e., IgG and IgM) and antibodies that bind to leukocyte Fc receptors (some subclasses of IgG) induce the pathologic lesions of immune complex disorders. Because IgA can activate complement by the alternative pathway, IgA-containing complexes may also induce tissue injury.

The important role of complement in the pathogenesis of the tissue injury is supported by the observations that during the active phase of the disease, consumption of complement decreases the serum levels, and experimental depletion of complement greatly reduces the severity of the lesions.

METHODS OF INCREASING GRAFT SURVIVAL

Because HLA antigens are the major targets of transplant rejection, minimising the HLA disparity between the donor and the recipient would be expected to improve graft survival. In the case of related donor kidney transplants, a beneficial effect of matching for class I HLA alleles has been observed. In cadaver renal transplants, matching for HLA class I antigens (HLA-A and HLA-B) has at best a modest effect on graft acceptance. Additional matching for class II antigens (HLA-DR) results in a definite improvement in graft survival. However, even HLA-matched unrelated donors are likely to differ from the host at one or more minor histocompatibility antigens. These antigens are formed by peptides derived from polymorphic proteins other than those encoded in the HLA complex. They evoke a weak or slower rejection reaction that, nevertheless, necessitates the use of immunosuppression.

Except in the case of identical twins, who are obviously matched for all possible histocompatibility antigens, *immunosuppressive therapy* is a practical necessity in all other donor-recipient combinations. The mainstay of immunosuppression is the drug *cyclosporine*. Cyclosporine works by blocking activation of a transcription factor called nuclear factor of activated T cells, which is required for transcription of cytokine genes, in particular, the gene for IL-2. Additional drugs that are used to treat rejection include azathioprine (which inhibits leukocyte development from bone marrow precursors), steroids (which block inflammation), rapamycin and mycophenolate mofetil (both of which inhibit lymphocyte proliferation), and monoclonal anti-T-cell antibodies (e.g., monoclonal anti-CD3 and antibodies against the IL-2 receptor α chain, which block T-cell activation and may opsonise and help to eliminate the cells).

Although immunosuppression has produced significant gains in terms of graft survival, immunosuppressive therapy carries its own risks. The price paid in the form of increased susceptibility to opportunistic fungal, viral, and other infections is not small. These patients are also at increased risk for developing EBV-induced lymphomas, human papillomavirus-induced squamous cell carcinomas, and Kaposi sarcoma.

To circumvent the untoward effects of immunosuppression, much effort is being devoted to induce donor-specific tolerance in host T cells. One strategy being pursued in experimental animals is to prevent host T cells from receiving costimulatory signals from dendritic cells during the initial phase of sensitisation. This can be accomplished by interrupting the interaction between the B7 molecules on the dendritic cells of the graft donor with the CD28 receptors on host T cells, as, for example, by administration of proteins that bind to B7 costimulators. The second signal for T-cell activation and renders the T cells anergic or induces their apoptosis. Antibodies that block CD40 ligand are also being tried; they presumably work by inhibiting humoral immune responses and CTL generation. In addition, giving donor cells to graft recipients may prevent reactions to the graft, perhaps because the donor inoculum contains cells, such as immature dendritic cells, that induce tolerance to the donor alloantigens. This approach may result in long-term *mixed chimerism*, in which the recipient lives with the injected donor cells. Such approaches are being tried in patients.

Transplantation of Other Solid Organs

In addition to the kidney, a variety of organs, such as the liver, heart, lungs, and pancreas, are also transplanted. Because patients in need of heart and liver transplants are severely ill, and because the transplanted heart or liver has to fit snugly into the space previously occupied by the host organ, the availability and size of the donor organ are of major importance, and take precedence over HLA match. Hence, HLA matching is not done in these cases. The rejection reaction against liver transplants is not as vigorous as might be expected from the degree of HLA disparity. The molecular basis of this "privilege" is not totally understood. Furthermore, with effective immunosuppression, the rejection reactions can be greatly reduced.

Transplantation of Hematopoietic Cells

Use of hematopoietic cell transplants for hematologic malignancies, certain nonhematologic cancers, aplastic anemias, and certain immunodeficiency states is increasing. Transplantation of genetically engineered hematopoietic stem cells is also likely to be useful for somatic cell gene therapy, and is being evaluated in some severe combined immunodeficiencies. Hematopoietic stem cells are usually obtained from the bone marrow but may also be harvested from peripheral blood after they are mobilised from the bone marrow by administration of hematopoietic growth factors. Several features distinguish bone marrow transplants from solid organ transplants. In most of the conditions in which bone marrow transplantation is indicated, the recipient is irradiated with lethal doses either to destroy the malignant cells (e.g., leukemias) or to create a graft bed (aplastic anemias). Three major problems arise in bone marrow transplantation: graft-versus-host (GVH) disease, transplant rejection, and immunodeficiency.

GVH disease occurs in any situation in which immunologically competent cells or their precursors are transplanted into immunologically crippled recipients, and the transferred cells recognise alloantigens in the host. GVH disease occurs most commonly in the setting of allogeneic bone marrow transplantation but may also follow transplantation of solid organs rich in lymphoid cells (e.g., the liver) or transfusion of unirradiated blood.

Recipients of bone marrow transplants are immunodeficient because of either their primary disease or prior treatment of the disease with drugs or irradiation. When such recipients receive normal bone marrow cells from allogeneic donors, the immunocompetent T cells present in the donor marrow recognise the recipient's HLA antigens as *foreign* and react against them. Both CD4+ and CD8+ T cells recognise and attack host tissues. In clinical practice, GVH disease can be so severe that bone marrow transplants are done only between HLA-matched donor and recipient. However, it is possible that with conventional HLA typing, subtle molecular differences between donor and recipient HLA molecules are not picked up, and these may be enough to trigger GVH reactions. For this reason, DNA sequencing methods are now being used for molecular typing of HLA alleles. Furthermore, HLA matching will miss minor histocompatibility differences, and these may also be enough to induce GVH disease.

Acute GVH disease occurs within days to weeks after allogeneic bone marrow transplantation. Although any organ may be affected, the major clinical manifestations result from involvement of the *immune system and epithelia of the skin, liver, and intestines*. Involvement of skin in GVH disease is manifested by a generalised rash leading to desquamation in severe cases. Destruction of small bile ducts gives rise to jaundice, and mucosal ulceration of the gut results in bloody diarrhea. Despite considerable tissue damage, the affected tissues are not heavily infiltrated by lymphocytes. It is believed that in addition to direct cytotoxicity by CD8+ T cells, considerable damage is inflicted by cytokines released by the sensitised donor T cells.

Immunodeficiency is a frequent accompaniment of GVH disease. The immunodeficiency may be a result of prior treatment, myeloablative preparation for the graft, a delay in repopulation of the recipient's immune system, and attack on the host's immune cells by grafted lymphocytes. Affected individuals are profoundly immunosuppressed and are easy prey to infections. Although many different types of organisms may infect patients, infection with cytomegalovirus is particularly important. This usually results from activation of previously silent infection. Cytomegalovirus-induced pneumonitis can be a fatal complication.

Chronic GVH disease may follow the acute syndrome or may occur insidiously. These patients have extensive cutaneous injury, with destruction of skin appendages and fibrosis of the dermis. The changes may resemble systemic sclerosis. Chronic liver disease manifested by cholestatic jaundice is also frequent. Damage to the gastrointestinal mucosa may cause esophageal strictures.

The immune system is devastated, with involution of the thymus and depletion of lymphocytes in the lymph nodes. Not surprisingly, the patients experience recurrent and life-threatening infections. Some patients develop manifestations of autoimmunity, postulated to result from the grafted CD4+ helper T cells reacting with host B cells and stimulating these cells, some of which may be capable of producing autoantibodies.

Because GVH disease is mediated by T lymphocytes contained in the donor bone marrow, depletion of donor T cells before transfusion virtually eliminates the disease. This protocol, however, has proved to be a mixed blessing: GVH disease is ameliorated, but the incidence of graft failures and the recurrence of disease in leukemic patients increases. It seems that the multifaceted T cells not only mediate GVH disease but also are required for engraftment of the transplanted marrow stem cells and control of leukemic cells. The latter, called *graft-versus-leukemia* effect, can be quite dramatic. Deliberate induction of graft-versus-leukemia effect by infusion of allogeneic T cells is being tested in the treatment of chronic myelogenous leukemia if the patient relapses after bone marrow transplantation.

The mechanisms responsible for rejection of allogeneic bone marrow transplants are poorly understood. It seems to be mediated by NK cells and T cells that survive in the irradiated host. NK cells react against allogeneic stem cells because the latter are lacking self-MHC class I molecules and hence fail to deliver the inhibitory signal to NK cells. There is great interest in exploiting the ability of NK cells to kill allogeneic hematopoietic cells for treating certain acute leukemias (another example of the graft-versus-leukemia effect). The hope is that allogeneic NK cells will kill the tumor cells but will not cause GVH reactions.

MORPHOLOGIC CONSEQUENCES OF IMMUNE COMPLEX

The morphologic consequences of immune complex injury are dominated by acute necrotizing vasculitis, with necrosis of the vessel wall and intense neutrophilic infiltration. The necrotic tissue and deposits of immune complexes, complement, and plasma protein produce a smudgy eosinophilic deposit that obscures the underlying cellular detail, an appearance termed **fibrinoid necrosis.**

When complexes are deposited in kidney glomeruli, the affected glomeruli are hypercellular because of swelling and proliferation of endothelial and mesangial cells, accompanied by neutrophilic and monocytic infiltration. The complexes can be seen on immunofluorescence microscopy as granular lumpy deposits of immunoglobulin and complement and on electron microscopy as electron-dense deposits along the glomerular basement membrane.

If the disease results from a single large exposure to antigen (e.g., acute serum sickness and perhaps acute poststreptococcal glomerulonephritis), the lesions tend to resolve, owing to catabolism of the immune complexes. A *chronic form of serum sickness* results from repeated or prolonged exposure to an antigen.

Continuous antigenemia is necessary for the development of chronic immune complex disease because, as stated earlier, complexes in antigen excess are the ones most likely to be deposited in vascular beds. This occurs in several human diseases, such as systemic lupus erythematosus (SLE), which is associated with persistent antibody responses to autoantigens. In many diseases, however, the morphologic changes and other findings suggest immune complex deposition but the inciting antigens are unknown. Included in this category are membranous glomerulonephritis, many cases of polyarteritis nodosa, and several other vasculitides.

CELL-MEDIATED (TYPE IV) HYPERSENSITIVITY

The cell-mediated type of hypersensitivity is initiated by antigen-activated (sensitised) T lymphocytes. It includes the *delayed type hypersensitivity reactions* mediated by CD4+ T cells, and *direct cell cytotoxicity* mediated by CD8+ T cells. It is the principal pattern of immunologic response not only to a variety of intracellular microbiologic agents, such as *Mycobacterium tuberculosis*, but also to many viruses, fungi, protozoa, and parasites. So-called contact skin sensitivity to chemical agents and graft rejection are other instances of cell-mediated reactions. In addition, many autoimmune diseases are now known to be caused by T cell-mediated reactions. The two forms of T cell-mediated hypersensitivity are described next.

Delayed Type Hypersensitivity

The classic example of delayed hypersensitivity is the *tuberculin reaction*, which is produced by the intracutaneous injection of tuberculin, a protein-lipopolysaccharide component of the tubercle bacillus. In a previously sensitised individual, reddening and induration of the site appear in 8 to 12 hours, reach a peak in 24 to 72 hours, and thereafter slowly subside. Morphologically, delayed type hypersensitivity is characterised by the accumulation of mononuclear cells around small veins and venules, producing a perivascular "cuffing".

There is an associated increased microvascular permeability caused by mechanisms similar to those in other forms of inflammation. Not unexpectedly, plasma proteins escape, giving rise to dermal edema and deposition of fibrin in the interstitium.

The latter appears to be the main cause of induration, which is characteristic of delayed hypersensitivity skin lesions. In fully developed lesions, the lymphocyte-cuffed venules show marked endothelial hypertrophy and, in some cases, hyperplasia. Immunoperoxidase staining of the lesions reveals a preponderance of CD4+ (helper) T lymphocytes.

With certain persistent or nondegradable antigens, such as tubercle bacilli colonising the lungs or other tissues, the initial perivascular lymphocytic infiltrate is replaced by macrophages over a period of 2 or 3 weeks. The accumulated macrophages often undergo a morphologic transformation into epithelium-like cells and are then referred to as *epithelioid cells*. A microscopic aggregation of epithelioid cells, usually surrounded by a collar of lymphocytes, is referred to as a *granuloma*. This pattern of inflammation that is sometimes seen in type IV hypersensitivity is called *granulomatous inflammation*.

The sequence of cellular events in delayed hypersensitivity can be exemplified by the tuberculin reaction. When an individual is first exposed to protein antigens of tubercle bacilli, naive CD4+ T cells recognise peptides derived from these antigens in association with class II molecules on the surface of antigen-presenting cells. This initial encounter drives the differentiation of naive CD4+ T cells to T_H 1 cells. The induction of T_H 1 cells is of central importance because the expression of delayed hypersensitivity depends in large part on the cytokines secreted by T_H 1 cells. Why certain antigens preferentially induce the T_H 1 response is not entirely clear, but the cytokine milieu in which naive CD4+ T cells are activated seems to be relevant.

Some of the T_H 1 cells enter the circulation and may remain in the memory pool of T cells for long periods, sometimes years. On intracutaneous injection of tuberculin in an individual previously exposed to tubercle bacilli, the memory T_H 1 cells recognise the antigen displayed by antigen-presenting cells and are activated. These T_H 1 cells secrete cytokines, mainly IFN-a, which are responsible for the expression of delayed-type hypersensitivity. Cytokines most relevant to this reaction and their actions are as follows:

- IL-12, a cytokine produced by macrophages and dendritic cells, is critical for the induction of the T_H 1 response and hence delayed hypersensitivity. On initial encounter with a microbe, the macrophages and dendritic cells that are presenting microbial antigens secrete IL-12, which drives the differentiation of naive CD4+ helper cells to T_H 1 cells. These, in turn, produce other cytokines. IL-12 is also a potent inducer of IFN-γ secretion by T cells and NK cells. IFN-α further augments the differentiation of T_H 1 cells.
- IFN-γ has many effects and is the key mediator of delayed-type hypersensitivity. Most importantly, it is a powerful activator of macrophages. Activated macrophages are altered in several ways: their ability to phagocytose and kill microorganisms is markedly augmented; they express more class II molecules on the surface, thus facilitating further antigen presentation; they secrete several polypeptide growth factors, such as platelet-derived growth factor (PDGF), which stimulate fibroblast proliferation and augment collagen synthesis; they secrete TNF, IL-1, and chemokines, which promote inflammation; and they produce more IL-12, thereby amplifying the T_H 1 response. Thus, activated macrophages serve to eliminate the offending antigen; if the activation is sustained, continued inflammation and, ultimately, fibrosis result.
- IL-2 causes autocrine and paracrine proliferation of T cells, which accumulate at sites of delayed hypersensitivity; included in this infiltrate are some antigen-specific CD4+ T_H 1 cells and many more bystander T cells that are recruited to the site.
- TNF and lymphotoxin are two cytokines that exert important effects on endothelial cells: (1) increased secretion of prostacyclin, which, in turn, favors increased blood flow by causing local vasodilation; (2) increased expression of P-E-selectins, adhesion molecules that promote attachment of the passing lymphocytes and monocytes; and (3) induction and secretion of chemokines such as IL-8. Together, all these changes in the endothelium facilitate the extravasation of lymphocytes and monocytes at the site of the delayed hypersensitivity reaction. The process by which T cells and monocytes exit the vasculature is generally similar to that described for neutrophils. The steps in this process are initial rolling on the endothelium, followed

by activation of integrins and firm adhesion and, ultimately, transmigration through the vessel wall.

- Chemokines produced by the T cells and macrophages recruit more leukocytes into the reaction site. This type of inflammation is sometimes called "immune inflammation."

T cell-mediated hypersensitivity is a major mechanism of defence against a variety of intracellular pathogens, including mycobacteria, fungi, and certain parasites, and is also involved in transplant rejection and tumor immunity. In addition to its beneficial, protective role, delayed type hypersensitivity can also be a cause of disease. *Contact dermatitis* is a common example of tissue injury resulting from delayed hypersensitivity. It may be evoked by coming in contact with urushiol, the antigenic component of poison ivy or poison oak, and manifests in the form of a vesicular dermatitis.

The basic mechanism is similar to that described for tuberculin sensitivity. On repeat exposure to the plants, the sensitised CD4+ cells of the T_H 1 type first accumulate in the dermis, then migrate towards the antigen within the epidermis.

Here they release cytokines that damage keratinocytes, causing separation of these cells and formation of an intraepidermal vesicle. Type I diabetes and multiple sclerosis are two diseases involving different organs in which tissue injury is caused by delayed type hypersensitivity reactions against autologous tissue antigens, mediated by the T_H 1 type of CD4+ T cells. In these examples of T_H 1-mediated autoimmune disease, there is some evidence that CD8+ cells may also be involved. In certain other forms of delayed hypersensitivity reactions, especially those that follow viral infections, cytokine-producing CD8+ cells may be the dominant effector cells.

Transplant Rejection

Transplant rejection is several of the immunologic reactions. A major barrier to transplantation is the process of *rejection*, in which the recipient's immune system recognises the graft as being foreign and attacks it. One of the important goals of present-day immunologic research is successful transplantation of tissues in humans without rejection.

Although the surgical expertise for the transplantation of skin, kidneys, heart, lungs, liver, spleen, bone marrow, and endocrine organs is now well in hand, it outpaces thus far the ability to confer on the recipient permanent acceptance of foreign grafts.

MECHANISMS CONCERNED IN REJECTION OF KIDNEY GRAFTS

The graft rejection depends on recognition by the host of the grafted tissue as foreign. The antigens responsible for such rejection in humans are those of the HLA system. Because HLA genes are highly polymorphic, any two individuals (other than identical twins) will express some HLA proteins that are different. Thus, every individual will recognise some HLA molecules in another individual as foreign (allogeneic) and will react against these. This reaction is the basis of rejection of grafts from one individual to another. *Rejection is a complex process in which both cell-mediated immunity and circulating antibodies play a role*; moreover, the relative contributions of these two mechanisms to rejection vary among grafts and are often reflected in the histologic features of the rejected organs.

T CELL-MEDIATED REACTIONS

The critical role of T cells in transplant rejection has been documented both in humans and in experimental animals. T cell-mediated graft rejection is called *cellular rejection*, and it is induced by two mechanisms: destruction of graft cells by CD8+ CTLs and delayed hypersensitivity reactions triggered by activated CD4+ helper cells. The recipient's T cells recognise antigens in the graft (the allogeneic antigens, or alloantigens) by two pathways, called *direct* and *indirect*.

- In the *direct pathway*, T cells of the transplant recipient recognise allogeneic (donor) MHC molecules on the surface of an antigen-presenting cell in the graft. It is believed that dendritic cells carried in the donor organs are the most important immunogens because they not only richly express class I and II HLA molecules but also are endowed with costimulatory molecules (e.g., B7-1 and B7-2). The T cells of the host encounter the dendritic cells either within the grafted organ or after the dendritic cells migrate to the draining lymph nodes. Both the CD4+ and the CD8+ T cells of the transplant recipient are involved in this reaction. CD8+ T cells recognise class I HLA antigens and differentiate into mature CTLs. This process of CTL differentiation is complex and incompletely understood. It appears to be dependent on the release of cytokines, such as IL-2, from CD4+ helper cells and CD40 ligand on the helper cells activating antigen-presenting cells to promote the differentiation of CTLs. Once mature CTLs are generated, they kill the grafted tissue by mechanisms. The CD4+ helper T-cell subset is triggered into proliferation and differentiation into T_H 1 effector cells by recognition of allogeneic class II molecules. As in delayed hypersensitivity reactions, cytokines secreted by the activated CD4+ T cells cause increased vascular permeability and local accumulation of mononuclear cells (lymphocytes and macrophages), and activate the macrophages, resulting in graft injury. The direct recognition of allogeneic MHC molecules seems paradoxical to the rules of self-MHC restriction: If T cells are normally restricted to recognising foreign peptides displayed by self-MHC molecules, why should these T cells recognise foreign MHC? Such recognition has been explained by assuming that allogeneic MHC molecules, with their bound peptides, resemble, or mimic, the self-MHC-foreign peptide complexes that are recognised by self-MHC-restricted T cells. The structural basis of such mimicry is not entirely clear.

- In the so-called *indirect pathway* of allorecognition, recipient T lymphocytes recognise antigens of the graft donor after they are presented by the recipient's own antigen-presenting cells. This process involves the uptake and processing of MHC molecules from the grafted organ by host antigen-presenting cells. The peptides derived from the donor tissue are presented by the host's own MHC molecules, like any other foreign peptides. Thus, the indirect pathway is similar to the physiologic processing and presentation of other foreign (e.g., microbial) antigens. The indirect pathway generates CD4+ T cells that enter the graft and recognise graft antigens being displayed by host antigen-presenting cells that have also entered the graft, and the result is a delayed hypersensitivity type of reaction. However, CD8+ CTLs that may be generated by the indirect pathway cannot directly recognise or kill graft cells, because these CTLs recognise graft antigens presented by the host's antigen-presenting cells. Therefore, when T cells react to a graft by the indirect pathway, the principal mechanism of cellular rejection may be T-cell cytokine production and delayed hypersensitivity. It is postulated that the direct pathway is the major pathway in acute cellular rejection, whereas the indirect pathway is more important in chronic rejection. However, this separation is by no means absolute.

Morphology of Rejection Reactions

On the basis of the morphology and the underlying mechanism, **rejection reactions are classified as hyperacute, acute, and chronic**. The morphologic changes in these patterns as they relate to renal transplants. Similar changes may occur in any other vascularised organ transplant.

Antibody-Mediated Reactions

Although there is little doubt that T cells are pivotal in the rejection of organ transplants, antibodies evoked against alloantigens in the graft can also mediate rejection. This process is called *humoral rejection*, and it can take two forms. *Hyperacute rejection occurs when preformed antidonor antibodies are present in the circulation of the recipient.* Such antibodies may be present in a recipient who has already rejected a kidney transplant. Multiparous women who develop anti-HLA antibodies against paternal antigens shed from the fetus may also have preformed antibodies to grafts taken from their husbands or children, or even from unrelated individuals who share HLA alleles with the husbands. Prior blood transfusions can also lead to presensitisation because platelets and white blood cells are rich in HLA antigens and donors and recipients are usually not HLA-identical.

When preformed antidonor antibodies are present, rejection occurs immediately after transplantation because the circulating antibodies react with and deposit rapidly on the vascular endothelium of the grafted organ. Complement fixation occurs, resulting in thrombosis of vessels in the graft, and ischemic death of the graft. With the current practice of cross-matching, that is, testing recipient's serum for antibodies against donor's cells, hyperacute rejection is no longer a significant clinical problem.

In recipients not previously sensitised to transplantation antigens, exposure to the class I and class II HLA antigens of the donor may evoke antibodies. The antibodies formed by the recipient may cause injury by several mechanisms, including complement-dependent cytotoxicity, inflammation, and antibody-dependent cell-mediated cytotoxicity. *The initial target of these antibodies in rejection appears to be the graft vasculature.* Thus, antibody-dependent, or *acute humoral rejection*, is usually manifested by a vasculitis, sometimes referred to as *rejection vasculitis*.

Acute Rejection

This may occur within days of transplantation in the untreated recipient or may appear suddenly months or even years later, after immunosuppression has been employed and terminated. As suggested earlier, acute graft rejection is a combined process in which both cellular and humoral tissue injuries contribute.

In any one patient, one or the other mechanism may predominate. Histologically, humoral rejection is associated with vasculitis, whereas cellular rejection is marked by an interstitial mononuclear cell infiltrate.

Acute cellular rejection is most commonly seen within the initial months after transplantation and is heralded by an elevation of serum creatinine levels followed by clinical signs of renal failure. Histologically, there may be extensive interstitial mononuclear cell infiltration and edema as well as mild interstitial hemorrhage.

As might be expected, immunoperoxidase staining reveals both CD4+ and CD8+ lymphocytes, and these cells express markers of activated T cells, such as the α chain of the IL-2 receptor. Glomerular and peritubular capillaries contain large numbers of mononuclear cells that may also invade the tubules, causing focal tubular necrosis.

In addition to causing tubular damage, CD8+ cells may also injure vascular endothelial cells, causing a so-called *endothelitis*. This form of cell-mediated vascular damage is limited to the endothelium and is distinct from the antibody-mediated vasculitis described later.

The affected vessels have swollen endothelial cells, and at places the lymphocytes can be seen between the endothelium and the vessel wall. The recognition of cellular rejection is important because, in the absence of an accompanying arteritis, patients promptly respond to immunosuppressive therapy.

Cyclosporine, a widely used immunosuppressive drug, is also nephrotoxic, and hence the histologic changes resulting from cyclosporine may be superimposed.

Hyperacute Rejection

This form of rejection occurs within minutes or hours after transplantation and can sometimes be recognised by the surgeon just after the graft vasculature is anastomosed to the recipient's. In contrast to the nonrejecting kidney graft, which rapidly regains a normal pink coloration and normal tissue turgor and promptly excretes urine, a hyperacutely rejecting kidney rapidly becomes cyanotic, mottled, and flaccid and may excrete a mere few drops of bloody urine. Immunoglobulin and complement are deposited in the vessel wall, and electron microscopy discloses early endothelial injury together with fibrin-platelet thrombi.

There is also a rapid accumulation of neutrophils within arterioles, glomeruli, and peritubular capillaries. These early lesions point to an antigen-antibody reaction at the level of vascular endothelium. Subsequently, these changes become diffuse and intense, the glomeruli undergo thrombotic occlusion of the capillaries, and fibrinoid necrosis occurs in arterial walls. The kidney cortex then undergoes outright infarction (necrosis), and such nonfunctioning kidneys have to be removed.

Chronic Rejection

In recent years, acute rejection has been significantly controlled by immunosuppressive therapy, and chronic rejection has emerged as an important cause of graft failure.

Patients with chronic rejection present clinically with a progressive rise in serum creatinine over a period of 4 to 6 months. Chronic rejection is dominated by vascular changes, interstitial fibrosis, and tubular atrophy with loss of renal parenchyma. The vascular changes consist of dense, obliterative intimal fibrosis, principally in the cortical arteries.

These vascular lesions result in renal ischemia, manifested by glomerular loss, interstitial fibrosis and tubular atrophy, and shrinkage of the renal parenchyma.

The glomeruli may show duplication of basement membranes; this appearance is sometimes called chronic transplant glomerulopathy. Chronically rejecting kidneys usually have interstitial mononuclear cell infiltrates containing large numbers of plasma cells and numerous eosinophils.

2

Pathogens and Animal Health

The easiest and most frequent way to transmit a pathogen is for an infected animal to come into contact with a healthy animal. Most disease is transmitted nose-to-nose or by other direct contact with sick animals. New additions to flocks or herds are a very common way to introduce pathogens. Stray pets that wander onto your farm or land can be another source of pathogen transmission.

If a sick animal is allowed into a show or fair, it may infect other nearby animals. In addition to stray pets; wild animals, free-flying birds, and insects can be responsible for the introduction of a pathogen.

People Movement People, vehicles, equipment, and clothing that are in contact with an infected animal may be contaminated with pathogens. These contaminated articles include equipment, machinery, and other objects used for the transportation, care, and management of livestock and poultry. If pathogens can be spread from animal to animal at shows and fairs, does that mean that we cannot take an animal to the show? We can purchase animals and go to the fair, but you should isolate those animals when you return. The wisest decision is to choose shows and make livestock and poultry purchases from places where disease prevention is always given high priority.

PRACTICAL STEPS THAT IMPROVE ISOLATION

Isolation (a form of quarantine) prevents contact between animals within a controlled environment. The idea behind isolation is to prevent contact between healthy animals and an animal that is or could be infected with a pathogen. Animals new to the herd or flock should be isolated away from the rest of the herd or flock for at least four weeks. Always isolate sick animals and only return them to their original group when they've fully recovered. Livestock returning from fairs or shows should also be isolated, since they could have picked up a new pathogen at the show. After returning to the farm, you should isolate your show animals to avoid the possibility of infecting other animals on your farm. Ideally, this should be in a completely separate place to avoid close contact. At minimum it could be a separate pen in a different building or at least a separate corner of the barn. This may represent some extra work, but it can be very important. Isolate all purchased animals for four weeks. If they are incubating a serious disease, the signs (i.e., diarrhea, hard breathing, etc.) of that disease will likely be observed in that 4 week isolation period.

This gives you some time to do follow-up testing or give booster vaccinations if needed. Always purchase good, healthy animals. Do not buy sickly animals, even if they are being sold for a "cheap" price. Livestock or poultry at bargain prices can be expensive!

Since people can carry pathogens on their body or clothing, it is necessary to control the visitors that may come in contact with your livestock or poultry. Always make sure visitors wear clean boots and clothing. It is best if visitors come only when you are there to escort them around your farm. Locked gates and doors will keep people out when you're not around. Signs and notices help to alert visitors of the potential risk that they may pose. Perimeter fences will keep people from accidentally wandering onto your property. Question visitors about the farms or animals they have visited or have been near recently, and ask if they have visited a foreign country within the past 5 days.

There are ways to help keep out wild animals, free-flying birds, and insects. Although it is impossible to totally prevent all contact between wildlife and our livestock, we can make barnyards and surroundings unattractive to many of these species. Cutting the grass and weeds around the outside of buildings will help prevent rodents from entering. Keep grain spills or other potential sources of food cleaned up and unavailable to wildlife.

Clean up old board piles or woodpiles and inspect buildings for possible hiding or denning areas. Inspect the haymow or other protected areas for evidence that cats, raccoons, or other animals that are likely using the hay or the straw for nesting areas. Store feed where wild animals, including rodents and birds, cannot make contact with it. Dispose of all waste feed in a way that will not attract pests. Maintain barns and buildings in good repair making it more difficult for animals or birds to gain access. Keep doors and windows shut when not needed for ventilation. Place screens or netting on the windows.

LOCATION AND CONSTRUCTION OF BUILDINGS

When planning the location of a barn or building used to house your livestock and/or poultry, isolation is key. The buildings or pens should be constructed to promote isolation from outside sources of disease. The facility should be a substantial distance from road traffic. It is important to consider possible exposure and the distance from other flocks, herds, or populations of domestic or wild animals. It is especially crucial that there is no nose-to-nose or beak-to-beak contact with other animals.

There should not be any contact with drainage water or waste runoff from other animals. When planning the organisation of either a large- or small-scale animal operation, the risk of disease can be greatly reduced whenever you keep different species and ages of the same species well separated. This is because pathogens producing little or no disease in one species can produce significant disease in another species. Mild infections in older animals can produce severe disease in young animals.Disease susceptibility within the same species varies with age.

Because they have acquired resistance, older animals may shed pathogens without getting sick themselves. These same pathogens can be serious to younger animals. Salmonella and E. coli are good examples of pathogens that can spread and produce extra trouble in young or newborn animals when they are not well separated from older stock.

Various ways may be available to achieve species/age separation. They will be different depending on many factors such as the value of the animals, the economic and labour resources available to the herd/flock owner, and such other factors as farm size and marketing opportunities. They may include separate buildings, well-separated pens, different caretakers, or, ideally separate premises for different species and ages. Certain types of farm operations are able to ensure separation of ages by using the "all-in/all-out" management practice. All-in/all-out management allows only one age at a time on a premise. No new or younger animals are brought in until all of the older animals are removed and the barn has been cleaned. This allows for cleaning and disinfecting before a new group of animals is brought onto the premises. The farm environment can be a very comfortable place where pathogens can survive and multiply.

Animal buildings, barns and pastures, as well as litter and manure surfaces, are often damp environments favourable for a dangerous increase in pathogen populations. Depending on the number of animals, very large accumulations of manure and litter may develop and become areas where pathogens can survive and multiply for many weeks or months.

SANITATION

How can you avoid bringing your animals to pathogens? The best way is to maintain a clean, dry, and sanitary environment. Cleaning is very important in reducing pathogen levels. Cleanliness means freedom from dirt, filth, and debris. Although buildings and equipment can appear clean when they are not visibly dirty, many microscopic contaminating pathogens can continue to be present. They remain and survive by clinging to surfaces, and by hiding in tiny cracks and crevices. The use of a disinfectant kills these remaining pathogens Proper disinfection results in reduction of pathogens. Reducing the amount of pathogens around your animals will decrease the risk of disease. Disinfectants are chemical agents that kill pathogens on contact. However, it is important to clean before disinfecting so that the pathogens become fully exposed to the disinfectant.

Pre-cleaning is doubly important because a disinfectant can be used up or inactivated by dirt. Disinfection and thorough, prompt drying are the last steps in the sanitation process. Without quick, thorough drying, any pathogens surviving the effects of the disinfectant can re-multiply to high numbers.

Hygienic Manure

Animal body wastes (feces, urine, etc.) ordinarily accumulate right in the environment (surroundings) where the animals spend most of their lives. In humans, of course, this does not occur because of toilets, plumbing and sewage treatment/disposal systems. Since flush toilets are not practical for animals, this ordinarily unavoidable problem for animals is partially reduced when manure or litter is managed in a way that, at least lowers the level of exposure to potential pathogens. Accumulated manure must be managed to reduce its volume, its odour, and to kill pathogens and weed seeds. The essential parts of manure management include collection, transfer, and storage. Spreading manure onto fields is the most common method of disposal.

The sun will dry manure after it is spread onto fields which will kill many pathogens. A proper level of airflow over manure surfaces promotes drying and the decrease of many pathogen populations. In contrast, stagnant air contributes to the build- up of moisture, and an increase in pathogen populations. Chemical treatments can be beneficial in manure management. Raising the pH to at least 12 for 30 minutes will kill most of the microorganisms present. Lime is usually used to raise the pH. Treatments that produce anaerobic conditions (very low oxygen) are also used. Anaerobic lagoons take advantage of a natural process where manure is digested by beneficial anaerobic bacteria. In contrast, aerobic lagoons add oxygen to the manure.

The addition of oxygen allows more common bacteria to survive and multiply and for the bacteria to break down the waste material. These bacteria convert the manure into carbon dioxide, water, and more beneficial lagoon bacteria. In animal pens or holding areas, we want the bedding and floor dry. Management of areas where water spillage or water accumulation is highest is very important. Good drainage and proper ventilation of buildings will help reduce the dangers of dampness. The frequent removal of wet litter and/or bedding and a continuous, modest flow of air over manure/litter surfaces will help to produce the drying needed to suppress bacteria.

MULTI-HOST PATHOGENS OF ANIMALS

Pathogens are often characterised as specialists or generalists based on the number of different host species they infect, as well as the **phylogenetic relatedness** among hosts. **Host range** can be associated with several factors, including the geographic ranges of pathogens and hosts, host and pathogen phylogeny, and life history traits. Some studies also point to the influence of mode of transmission in determining the host specificity of different types of pathogens. In addition, a pathogen's host range can be specific to a certain strain or subtype of the pathogen, which is the case for various subtypes of influenza A. It is worth noting that the designation of a narrow or wide host range is relative. For example, White Nose Syndrome, a disease caused by a pathogenic fungus infecting several bat species, could be said to have a narrow host range because it only infects one class of animals. One the other hand, it could be considered to have a wide host range compared to a pathogen like *Plasmodium falciparum*, the protozoan that causes malarial illness and only infects humans.

The majority of pathogens of animals are generalists that infect multiple host species, referred to as **multi-host pathogens** or **multi-host parasites**. Some multi-host pathogens are maintained in a **sylvatic** transmission cycle where the pathogen is maintained completely in multiple wildlife species. Among domesticated animal species, roughly 77% of pathogens of livestock and 90% of pathogens of domestic carnivores are known to be multi-host pathogens. Over 60% of all known human pathogens are **zoonotic** , meaning they originate in animals but can cross-infect humans. In some cases, humans can go on to infect humans or other animals (e.g., plague), while in others (e.g., West Nile Virus) humans are **dead-end hosts**. In the latter case, the pathogen causes disease in an individual human but further transmission to other hosts or vectors does not occur.

PATHOGENS INFECT MULTIPLE HOST SPECIES

The ability to infect multiple host species is not limited to pathogens of a specific type (e.g., virus, bacteria, helminth) or pathogens employing a particular mode of transmission. The following are only a small subset of multi-host pathogens (or diseases caused by multi-host pathogens) listed by their mode of transmission:

- Close contact/direct transmission (including direct contact, airborne, aerosol, bite, or sexual transmission): SARS, rabies, monkeypox, influenza, hantavirus, herpes, SIV (simian immunodeficiency virus)
- Non-close/indirect transmission (including fomites, environmental transmission): cholera, avian influenza, anthrax, brucellosis
- Intermediate host: *Schistosomiasis*, *Dicrocoelium dendriticum*
- Vector-borne transmission: West Nile virus, Lyme disease, Chikungunya.

Depending on the mode of transmission, some pathogens are considered to be obligate multi-host pathogens or parasites; these include parasites with complex life cycles and vector-borne pathogens. Parasites that exhibit a complex life cycle require a definitive host for reproduction and one or more intermediate host species for growth and development. Vector-borne pathogens are transmitted between hosts by an intermediate organism, often an arthropod like mosquitoes or ticks, referred to as a vector. Several factors can enable a pathogen to infect multiple host species. For example, genetic change in the pathogen can occur through selection or through random mutations, allowing the pathogen to become better adapted to infect a new host species. It is generally believed that the higher a pathogen's mutation rate, the more genetically diverse it will be and therefore the more likely it is that the pathogen is a generalist.

For example, RNA viruses mutate about 300 times faster than DNA viruses, and directly transmitted RNA viruses of humans are more likely to be zoonotic than directly transmitted DNA viruses of humans. Host speciation is another mechanism by which a pathogen that originally infects an ancestral host comes to infect multiple new host species. Generally, pathogens tend to infect host species that are phylogenetically similar to each other because these host species share traits (e.g., immunologic, antigenic, or ecological similarities) that make them susceptible to the same pathogens. Conversely, more distantly related host species do not share as many traits, decreasing the chances that they will share pathogen species. Recently speciated hosts share genetic similarities, potentially allowing a pathogen to infect both species. Introductions of non-native hosts and pathogens can also result in the infection of a new host species by providing new opportunities for infectious contact between pathogens and naïve hosts.

Infecting a wide range of host species is one way in which a pathogen's chance of persistence is increased. The ability to infect multiple host species is not always adaptive, however, and several ecological trade-offs are associated with the benefit of a broad host range. For example, while single-host pathogens tend to evolve an intermediate level of virulence in their host, virulence evolution in multi-host pathogens is more complex. A multi-host pathogen could be highly virulent in one host while exhibiting low virulence in another. The optimal virulence in each host will depend on how each host contributes to pathogen fitness.

Another cost of infecting multiple host species is the degree to which a pathogen can adapt to a host's immune system. If a pathogen only infects one host species, the pathogen can evolve to become highly proficient at evading the immune system of that host.

In multi-host pathogens, however, an adaptation in one host species may be maladaptive in another host species. For example, many vector-borne pathogens are viruses, and thus are expected to have a great deal of genetic diversity due to high mutation rates. However, experimental research (i.e., serial passage experiments involving transmission between an invertebrate vector and a vertebrate host) has shown that viral genetic sequences are largely unchanged after multiple transmissions between very different species. Moreover, viruses that were experimentally allowed to transmit between members of only one species rapidly adapted to that species, with coinciding loss of fitness often observed in the bypassed species. This host alternation is, therefore, a potential constraint on the genetic diversity of multi-host pathogens.

INVASION AND POPULATION DYNAMICS OF MULTI-HOST PATHOGENS

The invasion of a naïve population of hosts and subsequent epidemiological dynamics of multi-host pathogens are inherently different from single host systems because multiple host species provide multiple invasion pathways as well as multiple transmission routes. That is, if infection is unsuccessful in one host species, the presence of another host species provides an alternative route for the pathogen to invade a community. Both invasion and persistence are related to a theoretical quantity, R_0, referred to as the basic reproductive number and defined as the number of secondary infections resulting from a single primary infection in a completely susceptible population. If $R_0>1$, then an introduced pathogen is likely to persist and may cause an epidemic in the host population. In a community comprised of multiple host species, R_0 may be greater than one for one species, but less than one for another species. In this case, the community composition would determine whether or not the pathogen will persist at the community level.

The form of transmission (i.e., density-dependent versus frequency-dependent) also has implications for population dynamics of multi-host pathogens. Single-host pathogens that rely on density-dependent transmission rarely drive their host to extinction because the host population will drop below a threshold size such that pathogen transmission can no longer be maintained. Infecting multiple host species, as well as exhibiting frequency-dependent transmission (e.g., sexually transmitted or vector-borne pathogens), increases the chance of pathogen-induced host extinction because the threshold density for pathogen persistence is eliminated.

In host-pathogen systems with multiple hosts, disease dynamics can also depend on the competence of each host species for harboring and transmitting the pathogen, as well as the relative frequency of transmission between host and vector species. Accordingly, the community composition of potential hosts can have a large effect on pathogen dynamics, especially when competence within the host community varies substantially. Specifically, theory suggests that, in multi-host vector-borne pathogen systems, more diverse host communities may reduce pathogen transmission by decreasing contacts between infected vectors and highly competent hosts compared with single-species host systems. This phenomenon, referred to as the dilution effect, has been studied primarily in the Lyme disease system of the Northeast U.S., but has also been demonstrated in other multi-host pathogen systems. Some scientists have argued that while empirical evidence exists for the dilution effect in several multi-host pathogen systems, the mechanism by which disease dilution is occurring is often unknown.

MULTI-HOST PATHOGENS IN A CHANGING CLIMATE

As many pathogens are associated with tropical or equatorial areas of the world, it has been suggested that increased temperatures accompanying climate change will lead to the emergence, re-emergence, or persistence of many more pathogens.

Changes in climate are predicted to lead to range expansion and range shifts of pathogens, their hosts, and their vectors, making precise climate-associated changes in disease dynamics difficult to predict. Adverse effects of climate warming have already been discovered in some multi-host pathogen systems, like chytridiomycosis outbreaks (a fungal infection caused by *Batrachochytrium dendrobatidis*) in amphibian communities. Severe declines in amphibian diversity have been linked to warmer temperatures, which are thought to increase the growth of the fungus. Changes in climate are also known to alter certain animal behaviors, like the timing or spatial course of migration, which has the potential to alter multi-host pathogen transmission by changing when and where pathogens and parasites encounter their hosts, affecting both the time and size of disease outbreaks.

Why should we study multi-host pathogens?

Multi-host pathogen systems are intrinsically complex, shaped by pathogen and host dynamics as well as evolutionary, environmental, and climatic interactions. Understanding multi-host pathogens from an ecological perspective provides a variety of potential applications. Multi-host pathogens can, for instance, affect organisms and ecological dynamics far outside their host range. Depending on their effect on a host species (e.g., high virulence/ mortality, behavioral modification, reduced fitness/reproduction), multi-host pathogens may regulate not only populations and communities of host species, but also predator, prey, or competitor populations.

Understanding the ecology and evolution of multi-host pathogens may also be important for species conservation and biodiversity preservation. Some species that are now declining due at least in part to multi-host pathogens include bird species infected by avian malaria in Hawaii and West Nile Virus in the continental U.S. , bat species in the U.S. infected with the pathogenic fungus (*Geomyces destructans*) that causes White-Nose Syndrome , and seals infected with phocine distemper virus in Europe. Understanding the ecology of multi-host pathogens, particularly zoonotic multi-host pathogens, can provide information needed for shaping human health policy and may contribute to outbreak detection and other warning systems, or be central to programmes aimed at preventing or reducing transmission and human infections by multi-host pathogens.

THREATS TO ANIMAL HEALTH

Animals may be injured or die from the attacks by something as large and dangerous as a coyote or wolf to something just as dangerous, but only as small, as a molecule. This chapter takes you on a voyage into a world of animal health dangers. This fascinating voyage ranges from dangers that are visible to the naked eye (predators), to some that are more easily seen with a hand lens (internal and external parasites), to some that can only be seen with a microscope (bacteria and viruses).

The voyage continues all the way on down to the sub-microscopic world of disease-producing molecules! We know that disease-causing pathogens that we cannot see threaten our livestock. In addition, there are also other threats that can be easily seen. Predators may easily be detected; however, many often appear at night or when people are not around. There are ways to help keep out predators that could potentially physically attack and harm your animals.

First, remove all easily accessible food supplies. This may be very difficult depending on the size of the farm and the amount of livestock feed on-hand.

Keeping feed bins in good repair and sealed off will help prevent predators from entering areas where feed is stored. Keeping feed bins and feeding areas clean is very important.

An accumulation of waste feed in and around feed bunks attracts animals. A second suggestion is to remove water supplies. This is even more difficult on most farms. Stock tanks often provide a constant water supply and are attractive to predators such as raccoons. Preventing any unnecessary pools of water will help. A third suggestion is to modify habitat and reduce access. Clearing brush and keeping weeds away from barns and buildings will help deter animals just like it will help deter rodents. Finally, you can trap or control predators. There are a variety of traps available to catch animals. Barns and buildings should be kept in good repair. Predators often enter barns though open doors, or cracks and holes that may exist. Closing doors and patching cracks and holes helps to reduce the problem. Wire mesh or screening on windows can also help.

EXTERNAL PARASITES

Small yet visible threats to livestock include external parasites such as ticks, flies, fleas, lice, mosquitoes, mites, grubs, etc. Parasites are a threat to livestock health just as microbial (invisible) pathogens are a threat. Parasites can transmit and spread microbial pathogens in addition to the harm and damage that they naturally cause by irritating animals and sapping their energy. The most common way to control external parasites is through the use of pesticides or insecticides. Pesticides kill and help control the parasite populations. Pesticides are applied as sprays, dips, pour-ons, dusts, injectables, pastes, boluses, etc. Ear tags impregnated with insecticide are commonly used in cattle. Pesticides are very effective however, pesticides alone will not control a threat such as flies. Keeping the environment clean and sanitary helps eliminate fly breeding areas. Manure management will help reduce fly breeding areas. Fly eggs and larvae in thinly spread manure are killed by drying and heat.

INTERNAL PARASITES

Livestock and poultry can be attacked by a variety of large and small worm-like parasites such as roundworms and tapeworms. These internal parasites mainly invade the digestive passages while some also infest an animal's breathing passages. Other parasites can go even deeper into the "donut" or animal body reaching various vital organs and body tissues. Most parasites enter an animal's body when the animal eats the egg (ova) or an early life stage of the parasite. These ova or intermediate life stages come from the adult parasite reproducing and living inside an animal. They are typically passed by way of animal droppings onto the ground, or into the bedding or litter. As a result, an effective preventive strategy for many internal parasites rests on keeping animals from eating feed or licking surfaces contaminated by animal waste. It is a wise practice to keep animal yards, pens and buildings or other concentration points as clean from urine and accumulated fecal material as is practicable. Various powerful chemicals are sometimes used to treat different types of infestation. Their improper or inappropriate use may produce more damage to your animals than would be done by the parasites alone. Consequently, it is best to seek guidance from your local veterinarian before you treat for worms. In the final analysis, prevention is often less expensive than treatment.

Bacteria and Viruses

Bacteria and viruses are visible only when they are magnified hundreds to many thousands of times. They are not only able to attack the skin and digestive tract or respiratory linings of an animal's body, but, are often able to devastate the entire body including such organs as the brain, heart, liver and spleen.

Their extremely small size makes it possible for these pathogens to survive long periods of times outside an animal's body. They can survive in fur, hair and feathers, in nasal and other discharges from a sick animal, and in animal urine and fecal droppings. Viruses survive, and bacteria can actually multiply, in animal bedding, litter, and manure. Many survive long periods of time in the tiny particles of dust or soil present in the farm environment. With many tiny, scattered hiding places they can easily be carried to a new farm or group of animals riding on a person's clothing, on the surfaces of boxes, crates or equipment, and on the wheels of cars and trucks. Bacteria are single-celled microorganisms. They may live free in the environment or within a living cell. E. coli (Escherichia coli), Streptococcus, Staphylococcus, and Salmonella are a few examples of bacteria that can cause disease.

Viruses are tiny organisms that only grow inside the cells composing the animal's body. All viruses rely on a live animal host to reproduce. Examples of viruses that affect humans include the common cold (rhinovirus) or the flu (influenza) virus. Viruses can infect animals and cause respiratory symptoms (influenza viruses), diarrhea (rotavirus and coronaviruses), and numerous other disorders.

Although common viral or bacterial pathogens are unwelcome visitors or residents, certain ones can be especially troublesome. These special pathogens produce unusually high losses and/or unusual behaviour such as extensive tenderfootedness, incoordination, or slobbering. Milk or egg production may cease or drop sharply.

Your daily activities in caring for and feeding your animals put you in an excellent position where you could be among the first to spot the possible emergence of an especially unwelcome pathogen. As previously emphasised, isolation, traffic control, hygiene, and sanitation are the main ways to keep these essentially invisible pathogens from spreading to, and from reaching levels that can infect your animals! Medications, antibiotics, and vaccines are additional measures available to minimise the effects of bacterial or viral infections. Although powerful allies in an animal's battle with one or more pathogens, they are really the second line of defence.

The first line consists of all the steps you take to keep these microbes out or their numbers down to begin with. Again, those primary steps are herd/flock isolation, traffic control, hygiene, and sanitation.

Dangerous Molecules Prion

A prion is a very unusual infectious agent that is capable of causing an infection or disease. It is believed to be a self-reproducing protein structure that is similar to a virus. Prions cause prion diseases such as transmissible spongiform encephalopathies (TSE's). Unlike other infectious agents, prions produce no body-defending immune response. Mold Toxins. Mold growth takes place in feeds when they are damp. Mold does not always mean that you cannot use the feed.

Most molds are not toxic when fed to livestock; however, some are very toxic. The toxic by-products produced by molds (fungi) are known as mycotoxins (mold toxins). Mold toxins are very diverse because they are produced by many different molds.

Fungi that have the ability to produce mycotoxins are very common in grain and other livestock feeds and in the facilities and equipment used for transportation and storage of feed. They can be very dangerous causing reduced growth rates, lowered immunities, and increased susceptibility to various infectious diseases. Many feed companies are working on ways to sample and test feeds and grains for mold toxins before their sale.

Animals feeds must be kept dry. We must realise that feeding any moldy feedstuffs is a risk to the health of our animal.

There is less risk in feeding moldy feeds to fattening animals than for lactating or pregnant animals. The risk is also different among different species of animals. Dryness is the simplest way to control the growth of mold. Take the following steps when evidence of mold growth (musty odour, visible mold) is suspected.

- Stop the moisture source (fix the leak) correct condensation problems in outside storage bins.
- Thoroughly dry or discard all porous items.
- Scrub mold off hard surfaces with detergent and water, and thoroughly dry. Chlorine-containing products are of help in killing mold spores.

Bacterial Toxins

Many bacterialpathogens have the ability to produce toxins which add to the disease they produce, thus making them a double threat to your animals. Some toxin-producing bacteria form their poisons outside the body (botulism), while others do so inside the body (tetanus, black leg, and gangrene). These toxic bacteria can present a much greater risk than many ordinary bacteria. Vaccines are used to aid in their prevention, particularly where risks of exposure or infection are high.

Safeguarding Animal Health

The position of a flock or herd is not unlike that of a populous city of which the public health largely depends on the functions of an intelligent "health officer." The owner, should function as a health officer to his/her flock or herd. This element of animal management completes the picture of what it takes to keep animals safe and healthy. The health officer idea is especially important to many young people. After all, in the next few years, they may become the owners or managers of even larger populations of animals.

They need to be alert to threats to the health of their own animals, and to those belonging to others. They should be quick to recognise and properly respond to unusual diseases and to oversee and manage their own operations to prevent or minimise the effects of common diseases. The search for and application of disease prevention knowledge can be an enjoyable, enriching lifelong experience. An alert, active disease prevention "mind set" provides many dividends! By protecting your own animals from pathogens you are also helping to protect the animals of others.

3

Healing and Inflammation System

INTRODUCTION

Although there are many types of wounds, most undergo similar stages in healing that are mediated by cytokines and other chemotactic factors within the tissue. The duration of each state varies with the wound type, management, microbiologic, and other physiologic factors. There are 3 major stages of wound healing after a full-thickness skin wound.

Inflammation is the first stage of wound healing. It can be divided into several phases, resulting in the control of bleeding and the resolution of infection. During the initial phase, vasoconstriction occurs immediately to control hemorrhage, followed within minutes by vasodilation. During the second phase, cells adhere to the vascular endothelium. Within 30 min, leukocytes migrate through the vascular basement membrane into the newly created wound. Initially, neutrophils predominate (as in the peripheral blood); later, the neutrophils die off and monocytes become the predominant cell type in the wound. Debridement is the next phase of wound healing. Although neutrophils phagocytose bacteria, monocytes, rather than neutrophils, are considered essential for wound healing. After migration out of the blood vessels, monocytes are considered macrophages, which then phagocytose necrotic debris. Macrophages also attract mesenchymal cells by an undefined mechanism. Finally, mononuclear cells coalesce to form multinucleated giant cells in chronic inflammation. Lymphocytes may also be present in the wound and contribute to the immunologic response to foreign debris.

Proliferation is the second stage of wound healing. It consists of fibroblast, capillary, and epithelial proliferation phases. During the proliferation stage, mesenchymal cells transform into fibroblasts, which lay fibrin strands to act as a framework for cellular migration. In a healthy wound, fibroblasts begin to appear <“3 days after the initial injury. These fibroblasts initially secrete ground substance and later collagen. The early collagen secretion results in an initial rapid increase in wound strength, which continues to increase more slowly as the collagen fibers reorganise according to the stress on the wound.

Migrating capillaries deliver a blood supply to the wound. The center of the wound is an area of low oxygen tension that attracts capillaries following the oxygen gradient. Because of the need for oxygen, fibroblast activity depends on the rate of capillary development. As capillaries and fibroblasts proliferate, granulation tissue is produced. Because of the extensive capillary invasion, granulation tissue is both very friable and resistant to infection.

Epithelial cell migration begins within hours of the initial wound. Basal epithelial cells flatten and migrate across the open wound. The epithelial cells may slide across the defect in small groups, or "leapfrog" across one another to cover the defect. Migrating epithelial cells secrete mediators, such as transforming growth factors α and β, which enhance wound closure. Although epithelial cells migrate in random directions, migration stops when contact is made with other epithelial cells on all sides (ie, contact inhibition). Epithelial cells migrate across the open wound and can cover a properly closed surgical incision within 48 hr.

In an open wound, epithelial cells must have a healthy bed of granulation tissue to cross. Epithelialisation is retarded in a desiccated wound. Remodeling is the final stage of wound healing. During this period, the newly laid collagen fibers and fibroblasts reorganise along lines of tension. Fibers in a nonfunctional orientation are replaced by functional fibers. This process allows wound strength to increase slowly over a long period (up to 2 yr). Most wounds remain 15–20% weaker than the original tissue. However, the urinary bladder and bone regain 100% of their original strength after wounding and repair.

SIGNALS FOR SWITCHING FROM KILLING TO HEALING

A crucial commitment made late in inflammation is to convert the response from the antibacterial, tissue-damaging mode to a mode that promotes tissue repair and epithelial closure. The timing is critical — to close a wound before it is disinfected invites disaster. Some of the signals involved are revealed by the failure of mice to resolve late-phase inflammation when they are deficient in the CD44 hyaluronan receptor, secretory leukocyte protease inhibitor (SLPI) or TNF.

Continuing from the point reached in the description, as long as microbial and host pro-inflammatory stimuli predominate, macrophage-derived chemokines continue to attract neutrophils. ROI and hyaluronidase from macrophages and neutrophils break down hyaluronic acid in the extracellular matrix to low molecular weight fragments. Like the heat-shock proteins, HMGB1 protein and *N*-formyl peptides described earlier, hyaluronan fragments act as signals of injury, working via CD44 on macrophages to induce the further release of chemokines and perhaps MMPs. Neutrophils with engaged integrins are activated by macrophage-derived TNF to release abundant elastase. Elastase and ROI activate MMPs. MMPs activate macrophage-derived latent transforming growth factor-β (TGF-β), the most potent known chemoattractant for neutrophils. MMPs also degrade collagen, proteoglycans and fibronectin. Elastase degrades latent TGF--binding protein, contributing to the activation of TGF>β.

The transformation from tissue damage to tissue repair begins as complement, neutrophils and macrophages kill microbes, and macrophages secrete more SLPI, a serine protease inhibitor expressed late after exposure to microbial products or cytokines. SLPI has anti-inflammatory and wound-healing effects that include suppressing the release of elastase and ROI by TNF-stimulated neutrophils, inhibiting elastase that has already been released and preventing the breakdown of TGF-β. Furthermore, SLPI binds and synergises with proepithelin, a cytokine that promotes epithelial growth and suppresses neutrophil activation, protecting it from proteolytic conversion into pro-inflammatory epithelins. CD44-positive macrophages clear the hyaluronan fragments.

Fresh neutrophils no longer enter the site, and those present undergo apoptosis. Macrophages ingest apoptotic neutrophils and degrade their residual stores of elastase. TNF induces macrophages to release interleukin-12, which induces lymphocytes to release interferon-γ (IFN-γ). IFN-γ acts early on to induce macrophage chemokine production, but now suppresses it. Ingestion of apoptotic neutrophils elicits more TGF-β from macrophages, and the predominant action of TGF-β is no longer the recruitment of neutrophils, but instead the promotion of tissue repair. Thus, TNF, IFN-γ and TGF-β join PGE2 as examples of molecules whose actions switch from pro-inflammatory to anti-inflammatory, depending on timing and context.

suppress inflammation

ROIs and RNIs are two additional sets of molecules that can either promote or suppress inflammation. Chronic granulomatous disease (CGD), a genetic disorder predisposing to life-threatening bacterial and fungal infections, results from a deficiency in the ROI-producing enzyme phagocyte oxidase (phox). In CGD, chronic inflammation sometimes seems to precede infection or long outlast it.

The clinical impression of an exaggerated granulomatous response in CGD has been confirmed in phox-deficient mice, which form abnormally large granulomas when injected with sterile fungal cell walls. These observations demonstrate that phox has an important anti-inflammatory role, such as oxidatively inactivating chemotactic factors, even though phox can be profoundly pro-inflammatory by virtue of oxidising tissue constituents, oxidatively activating metalloproteinases and oxidatively inactivating protease inhibitors. Similarly, mice deficient in inducible nitric oxidase synthase (iNOS) display a triple phenotype — increased susceptibility to infection, reduced inflammation or excessive inflammation — depending on the experimental setting. Without infection or other experimental intervention, however, mice lacking phox or iNOS appear normal.

GENES WHOSE DISRUPTION PREDISPOSES TO INFLAMMATION

Another level of control is revealed by the fact that there are numerous genes whose disruption predisposes to inflammation in people or mice living under conventional conditions without evident provocations that are known to elicit inflammation in wild-type hosts.

That loss-of-function mutations can lead to spontaneous inflammation was probably shown first and most clearly by human C1q deficiency. This disorder confers a >90% incidence of systemic lupus erythematosus. Although incomplete, includes over 50 genes implies that health does not arise passively from a lack of pro-inflammatory stimuli. On the contrary, potentially inflammatory stimuli seem to be ubiquitous, and it takes an active process to avoid over-reacting to those that pose a minimal threat.

The diverse genes necessary to suppress spontaneous inflammation can be classified into functional sets. Although such groupings are subjective, it is difficult to posit less than three key elements of the tonic anti-inflammatory state. First is the solubilisation and clearance of immune complexes and cellular debris. The second element is the balanced progression of leukocytes and lymphocytes through programmes of activation, proliferation and apoptosis. Third is the avoidance of oxidative injury, such as by disposal of haem and by constitutive restraint on the respiratory burst activity of macrophages that are continually exposed to particulate stimuli.

Gene products are demonstrated importance for avoiding inflammation. In a paradox that is now familiar, most of these proteins, such as TNF-R1 and NF-B, are better known for their essential contributions to promoting inflammation. Again it emerges that pro-inflammatory gene products are often essential effectors of anti-inflammatory homeostasis.

Many of the phenotypes are strongly dependent on genetic background, age, sex and environmental conditions (such as intestinal flora). The profound influence of epistatic or nongenetic factors is apparent when considering the following contrasts. First, mice of one strain contrasted with mice of another, such as those whose deficiency in interleukin-1 receptor antagonist (IL-1Ra) leads either to arthritis or arteritis. Second, one person contrasted with another, such as those with NOD2 mutations who develop either enterocolitis or arthritis (NOD2 is an intracellular lipopolysaccharide- and kinase-binding protein with a caspase-recruitment domain (CARD), a nucleotide-binding domain and leucine-rich repeats (LRRs)). Third, people contrasted with mice, such as those whose disrupted C4 gene predisposes to rash or glomerulonephritis, respectively.

It is a puzzle that disrupting a given gene has tissue-specific consequences when expression of the gene in question is not restricted to, or in some cases even manifest by, the tissue that is inflamed. This is not fully explained by the ability of leukocytes and lymphocytes to enter any tissue. Overall, the sites most frequently affected by inflammation in association with the listed mutations (skinlung>kidney>joints> colonliver> heart> pancreas eyes>other organs) are those that are anatomically large (skin, lung, colon and liver), continually exposed to microbes (skin, lung, colon and conjunctivae), or prone to trapping immune complexes (kidney, joints and skin).

Finally, collection of mechanistic mysteries that invite investigation. Among the most intriguing are the genes whose mutation predisposes to periodic fever syndromes. Cryopyrin resembles NOD2, Toll-like receptors and CD14 in carrying an LRR. Pyrin and cryopyrin contain pyrin domains, which are predicted to share structural features with CARDs, such as those present in NOD2. What is the mechanism of the anti-inflammatory actions of these proteins? Do they have pro-inflammatory actions as well? What are the contributions of their LRR, pyrin and CARD domains? Why do the phenotypes of their mutations mimic those associated with mutations in genes encoding two additional and very different proteins, TNF-R1 and mevalonate kinase? Mevalonate kinase is essential for synthesis of isoprenoids and cholesterol. Perhaps this sheds light on the unexplained anti-inflammatory effects of statins, drugs that block cholesterol synthesis at a later step. Administration of statins might lead to accumulation of a cholesterol precursor whose formation depends on mevalonate kinase. Perhaps this intermediate has potent anti-inflammatory actions on its own, or confers such actions on a protein to which it becomes attached.

Perspective

Our bodies sustain the replication of hundreds of different genomes. Only the largest is heritable in the germ line. To preserve its opportunity to be transmitted, the germ line genome must encode hair-trigger vigilance against take-over of the soma by genomes that replicate far faster. The rapid mobilisation of microbicidal defences evolved at the cost of potentially suicidal autotoxicity.

Thus, for host survival, two sets of mechanisms must be matched: the ability to mount a rapid inflammatory response to injurious microbial invasion, and the ability to refrain from doing so otherwise. For those seeking the origins of inflammatory or autoimmune diseases, this analysis encourages two lines of inquiry: First, what might predispose to the formation, modification or relocalisation of endogenous molecules such that they activate detection systems that normally report injury and infection? Second, are there dysfunctions in pathways whose integrity is required to prevent inflammation from arising spontaneously?

For those trying to promote inflammation, the analysis offered above commends combinations of signals that mimic both injury of the host and the presence of infectious agents, without resorting to either.

For those developing anti-inflammatory therapies, the need for each of several go signals suggests that it should be relatively straightforward to interrupt inflammation. Unfortunately, the redundancy of many signals (over-determination) complicates this goal. The recognition of stop signals offers additional opportunities to abort inflammation. However, predictability is complicated by the tendency of signals to shift sense, as illustrated for PGE2, TNF, IFN-, TGF-, ROIs and RNIs. Finally, there remains the dilemma that the more broadly an agent suppresses inflammation, the more likely it will exacerbate infections. Corticosteroids taught this lesson, and TNF-neutralising agents reinforced it. Nonetheless, many of those working in anti-inflammatory research are optimistic. Experimental biology is uncovering an unprecedented wealth of molecular detail at a time when systems biology seems poised to put into perspective the complexity and dynamics of the inflammatory process. An alliance between experimental and systems biology should be a powerful force to identify points of control amenable to relatively safe and effective intervention.

REGENERATION IN ZEBRAFISH

The zebrafish has emerged as a powerful model organism for the application of genetics to study not only vertebrate development, but also regeneration. Among other tissues, zebrafish can regenerate retina, fins, and heart. It remains unknown whether zebrafish fin regeneration relies on "dedifferentiation" or stem/progenitor cell activation. Because cell implantation or tissue grafting is not yet an option in adult zebrafish, transgenesis has become the method of choice to address mechanisms of regeneration. The recent employment of transgenic lines during zebrafish heart regeneration clearly implicates a reserve progenitor cell and points the way for future experiments that should be designed to track down the origin of the blastema.

APPENDAGE (FIN) REGENERATION

While the zebrafish response to appendage amputation is very similar to that of amphibians, the anatomy of the fish fin differs significantly. Fish appendages are composed of multiple fin rays, each of which produces their own blastema. Wound healing occurs through a process of epithelial cell migration to form a wound epidermis, which thickens to form a layered structure containing a basal layer of distinctive cuboidal cells. Like urodele amphibians, the uninjured tissues in zebrafish close to the amputation plane become "disorganised", and cells of the mesenchyme proliferate, appear to migrate and give rise to a blastema (12–48 hpa, hours post amputation). The zebrafish blastema then compartmentalises at the onset of regenerative outgrowth (48 hpa), a period of rapid cell division, while differentiation and patterning conspire to rebuild the missing structures.

Presently, the heterogeneity of early and late blastemas is unknown, but appears to be a crucial aspect of early progenitor cell organisation. Cells of the early blastema are proliferative and the G2 phase of the cell cycle at this stage is greater than 6 hrs long, indicating that cells are cycling slowly. Cells in this region express markers including *msxb*, *msxc*, *sly1*, *mps1*, *fgf20a*, and *hsp60*. Thus, the early blastema may consist of a homogeneous cell population, or it may instead consist of a heterogeneous population of cells that differentially express these markers.

By the onset of regenerative outgrowth (48 hpa) the blastema begins to resolve into two distinct domains. Cells of the proximal blastema proliferate rapidly, achieving a G2 length of roughly 1 hr by 72 hpa, and express *mps1*, *hsp60*, and PCNA. Cells of the distal blastema do not proliferate and express *msxb*, *msxc*, and *sly1*.

In addition, the distal-most cells of the distal blastema region express *fgf20a* by 72 hpa, illustrating sub-regionalisation within the distal zone. It has been suggested that the distal zone is important to direct regenerative outgrowth and that the distal zone cells may be undifferentiated progenitor cells because they express *msxb* and *msxc*, which associate with undifferentiated cell types in other model organisms. Further experimentation will be necessary to validate or refute these possibilities.

BrdU pulse-chase experiments suggest that cells segregate to the distal blastema at the end of blastema formation around 36–48 hpa, and that these cells are descendents of formerly proliferating cells. However, it is unclear how this compartmentalisation occurs.

If the early blastema is composed of a homogeneous cell population, perhaps these cells are induced to restrict their expression profile at the onset of compartmentalisation. Alternatively, if the early blastema is a heterogeneous cell population in which each cell already expresses its position-specific profile, then perhaps compartmentalisation occurs by a process of cell sorting. How the shift from slow cycling to rapid cycling is controlled and how it is coordinated with compartmentalisation of the blastema remains unknown. New tools such as promoter driven cell ablation should help dissect the biological function of the distal and proximal domains.

Additionally, gene specific targeting combined with CreLox strategies should allow cell lineage tracing to definitively dissect the origin of the blastema in zebrafish appendages.

Functional genetics

Mutagenesis screens for temperature sensitive alleles that affect regeneration brought caudal fin regeneration to the forefront of regeneration research. Although these screens are limited by the fact that complete coverage of the genome with temperature sensitive alleles is not possible, genes associated with regeneration have been uncovered.

By combining forward genetics, pharmacology, and transgenic overexpression a potent toolbox is now available in zebrafish for a thorough investigation of its regenerative processes.

These strategies have already provided strains that are defective at various stages of regeneration. For example, fish harboring a mutation in either *fgf20a* or *hsp60* never form a blastema. Interestingly, *hsp60* is expressed in early mesenchymal blastema cells and appears to play a role in the mesenchymal cell response to nearby amputation. On the other hand, fish with a mutation in *mps1*, a mitotic checkpoint protein, exhibit normal wound healing and early blastema formation, but are defective in the regenerative outgrowth phase. Defects in *mps1* animals are observed in the proximal blastema zone at roughly 48 hpa during the transition to rapid cell proliferation. While *mps1* is clearly essential for proper regeneration, it sems to play a "house-keeping" type of role in rapidly proliferating cells.

In fish with a mutation in the *sly1* gene, wound healing and mesenchymal disorganisation are unafftected, but the proximal blastema zone is never established and the distal blastema zone is enlarged. Because the yeast *sly1* gene product is involved in protein trafficking and zebrafish *sly1* expressing cells segregate into the distal zone during normal blastema compartmentalisation, the *sly1* mutation has been hypothesised to cause defective signaling from the distal blastema cells to the more proximal proliferating cells. However, the nature of the presumed signal(s) remains unknown. While the *mps1* and *sly1* mutant phenotypes provide resources to study the compartmentalisation stage of zebrafish regeneration, the genomic regions surrounding these genes should also provide useful enhancer sequences for future investigations.

In fish, the FGF signaling pathway plays a very early role in establishing proper epidermal/mesenchymal interactions and blastema formation, just as suggested for FGF signaling during anuran amphibian regeneration. The FGF signaling pathway is essential for fin regeneration as evidenced using pharmacological, morpholino, and dominant negative transgenic approaches. Fish harboring an *fgf20a* loss of function mutation display an abnormally thickened wound epidermis, improper basal layer formation, failure of uninjured tissue near the amputation to disorganise, and absence of a blastema. *fgf20a* is expressed within the first 6 hours after amputation in the mesenchyme directly beneath the wound epidermis while the FGF receptor, *fgfr1*, is expressed by18 –24 hpa in fibroblast-like pre-blastemal mesenchymal cells just proximal and distal to the amputation plane. These combined data cement FGF signaling, either directly or indirectly, as an upstream regulator of stem/progenitor cell formation, migration, proliferation, or organisation. The cells expressing *fgfr1* may represent a fruitful target of future research focused on elucidating the nature of regenerative stem/progenitor cells in zebrafish.

Two other pathways, activin/TGF□ and Wnt, have also been implicated in the earliest steps of the regenerative response. Microarray experiments revealed that *activin-□A*, a member of the secreted TGF□ superfamily, is upregulated as early as 1 hpa. Manipulations that inhibit activin/TGF□ signaling cause an early block in the regenerative response by reducing mesenchymal disorganisation, *msxb* expression, and/or cell proliferation. On the other hand, inhibition of canonical

Wnt signaling (via overexpression of DKK or dominant negative TCF3) or activation of non-canonical Wnt signaling, blocks stratification of the wound epidermis and blastema formation. Wnt signaling inhibition also eliminates expression of both *fgf20a* and downstream FGF targets, suggesting that Wnt signaling is upstream of FGF signaling following amputation. Thus, regeneration of the zebrafish tail fin is akin to that of the anuran tadpole tail in that both systems utilize the same signaling pathways in apparently similar ways. These data provide a glimpse into the complex coordination of signaling pathways during the early regenerative response. Elucidation of the interactions between the basal layer of the wound epidermis and the responsive pre-blastemal stem/progenitor cells is paramount to understanding the vertebrate response to regeneration.

Unanswered questions

Naturally, great progress has extended old questions while raising a cohort of new ones:

- Do blastema cells derive from differentiated cells or undifferentiated reserve cells? While long term pulse-chase BrdU studies have apparently ruled out label retaining (slow cycling) progenitor cells, the presence of rapidly cycling stem/progenitor cells in the fin is suspected because there is a high steady-state level of cell proliferation in the fins of intact, unamputated animals.
- What is the source of the basal layer of the wound epithelium and how is it specified? *Lef1* appears in cells prior to the formation of a morphologically recognisable basal layer, but are they really presumptive cells of the basal layer? How are the multiple essential signaling pathways coordinated during this specification? The basal layer appears to be essential for regeneration, but why? Does it signal and induce blastema formation or is it simply a source of growth factors or extracellular matrix?
- What purpose is served by "disorganisation" of the uninjured mesenchymal tissue? Do these cells give rise to the proliferating blastemal cells, do they modify the extracellular environment to make it permissible for blastema activity, or both?
- What is the biological purpose of blastema compartmentalisation? Is the distal blastema a source of progenitor cells for the proximal blastema, or is it a source of secreted factors as implied by the *sly1* mutant phenotype?
- Given the ever-expanding toolkit available for the study of zebrafish biology, it is expected that answers to these questions will be forthcoming in the foreseeable future.

ORGAN (HEART) REGENERATION

Studies of zebrafish heart regeneration illustrate that careful lineage experiments are a pre-requisite for assessing the cellular source of regeneration. When adult fish hearts expressing red fluorescent protein (RFP) driven by a cardiomyocyte(CM)-specific promoter are transected, cells in the regenerating region express markers of cardiac embryonic progenitor cells and fluoresce red at a much reduced level compared to the surrounding differentiated CMs. This could be interpreted as the dedifferentiation of CMs, or as the differentiation of progenitor cells turning on the CM promoter for the first time. To distinguish between these possibilities, fish expressing both RFP and GFP from the same CM promoter were analysed. Because RFP folds and fluoresces slower than GFP and it is more stable, the appearance of RFP-/GFP+ cells after amputation would represent newly differentiating progenitor cells, while RFP+/GFP- cells would represent cells undergoing dedifferentiation. When the experiment was performed, these double transgenic hearts yielded RFP-/GFP+ cells throughout the regeneration process, indicating that regeneration of the myocardium results from the differentiation of progenitor cells and not from dedifferentiation.

While this provides important support for the progenitor hypothesis, the approach includes a key assumption: if a cell was to dedifferentiate, the cardiomyocyte promoter would turn off and proteins would undergo their normal turnover.

Because the potential mechanism used to accomplish dedifferentiation is unknown, it remains possible that a dedifferentiating cell would rapidly degrade a large number of protein products and the differences between RFP and GFP stability would no longer be observable.

Where do the progenitor cells come from? Using similar approaches during tissue growth in unamputated animals, recent work revealed that CM progenitor cells are induced throughout the entire adult myocardium during rapid homeostatic tissue growth. In addition, epicardial-derived cells from the outer edge of the heart actively migrate into the myocardium during both regeneration and homeostasis to build vascular tissue. Additional analysis of cell proliferation with more markers and at earlier time points will be required to fully address the origin of the progenitor cells. Nonetheless, the type of careful and creative experimental approaches used to dissect cardiac regeneration in zebrafish will be necessary in all vertebrate regenerative contexts to critically evaluate the contributions of dedifferentiation and stem/progenitor cells.

REGENERATION IN PLANARIANS

Planarians are bilaterally symmetric animals that possess derivatives of all three germ layers (endo-, ecto-, and mesoderm) and display astonishing regenerative abilities. Relative to cells involved in the vertebrate regeneration response, the source of regenerative cells in planarians is much less controversial. Planarians recruit an experimentally accessible population of adult stem cells called neoblasts that are distributed throughout the body. Classically defined by morphology, sensitivity to □-irradiation, and mesenchymal distribution, neoblasts are undifferentiated cells with a large nucleus and very little cytosol.

With the exception of the germline, neoblasts are thought to be the only planarian cells capable of division. After animals are exposed to □-irradiation, cell division ceases and neoblasts are lost, rendering the animals incapable of regeneration or homeostatic tissue turnover.

The loss of neoblasts is manifested by a characteristic ventral curling and the eventual lysis of the animals. In contrast to irradiation, neoblasts respond to amputation with a proliferative burst, which results in the formation of a regeneration blastema and the eventual restoration of the missing body parts. Because only injections of neoblast-enriched preparations can restore longevity and regeneration to irradiated animals, it is believed that in planarians, regeneration and tissue homeostasis are primarily driven by neoblast function.

Because small neoblast-containing fragments cut from almost any location in the adult animal can produce entire, properly proportioned planarians, it is thought that neoblasts are collectively totipotent. While the neoblasts are totipotent as a population, the differentiation potential of any given neoblast is unkown and the molecular nature of the neoblast population remains poorly described.

In recent years, the molecular dissection of planarian regeneration has been aided immensely by the discovery that RNA interference (RNAi) can be used to interrogate gene function. In one broad stroke, it was shown that over a thousand genes from an RNAi library could be screened for phenotypes in a relatively short amount of time. This work identified 240 genes associated with regeneration defects. From this collection, 140 gene perturbations blocked, limited, or reduced regeneration, 48 of which caused the characteristic curling/lysis phenotype observed after irradiation, indicating that neoblast function was compromised.

In addition to RNAi, sequencing of the planarian genome, generation of EST libraries, production of antibodies, development of cell-specific whole mount in situ hybridisation, and fluorescence in situ hybridisation (FISH) methods, labeling of proliferative cells, development of fluorescence activated cell sorting (FACS) protocols, and single cell RT-PCR techniques have catapulted planarians from an academic curiosity to a viable and fascinating model for regeneration research. Because the cells involved in planarian regeneration and homeostasis are identified and these cells can be studied *in vivo*, the set of questions one can currently address using planarians are very different from those being addressed by amphibians or zebrafish experimentation. These questions are focused on the biology of planarian stem cells and the mechanisms utilised to control their self-renewal, fate choice, and differentiation:

- Are all neoblasts the same or are there subsets of lineage-restricted cells?

- What is the extent of their molecular heterogeneity?
- How, when, and where do neoblasts choose their fate and decide to differentiate, especially in the context of regeneration?
- How is this decision controlled?
- Is differentiation immediate, or is it a slow process?
- Does it happen in stages?
- If it does happen in stages, what are the choices made at each step?

Heterogeneity and lineage of neoblasts

Researchers have begun to answer the above questions using the tools of modern molecular biology, but there is much more to learn. Studies using the DNA analog bromodeoxyuridine (BrdU) have shed light on the distribution and fate of neoblasts because this technique allows researchers to both visualise cells that are actively proliferating and follow those cells over time. After a single pulse of BrdU, only neoblasts are initially labeled, indicating that this population is rapidly dividing. Labeled cells are distributed throughout the parenchyma (mesenchyme) of the animal and are conspicuously absent from the pharynx and from a region anterior to the photoreceptors, the two regions of the animal that are unable to regenerate. This distribution fits the classical criteria for neoblast distribution. An antibody that recognises phosphorylated histone H3 (H3P), which marks dividing cells, reveals a similar neoblast distribution as do cell cycle specific probes.

If animals are fixed at various time points after BrdU is administered, labeling is observed in differentiated cell types roughly 35 hours after the BrdU pulse. Labeled cells can be observed in the post-mitotic epithelium several days later. When planarians are exposed to long-term continual doses of BrdU, all neoblasts, as defined by morphology, are eventually labeled. Finally, if a pulse of BrdU is administered prior to amputation, a large number of labeled cells are found in the regenerating tissue, showing that the progeny of neoblasts that were actively dividing prior to amputation make a major contribution to the regenerating tissue. Combined, these data argue that the neoblast population constantly divides to replace cells lost to turnover. In addition, the cells of regenerating tissue are derived from the dividing neoblasts.

The neoblast population likely contains both stem cells and their committed progeny. However, the heterogeneity of the neoblast population has been difficult to address because neoblasts are defined by their morphology. Evidence of heterogeneity among neoblasts has recently been suggested by electron microscopy and FACS experiments that associate chromatoid body-containing cells with what appears to be different stages of differentiation. Chromatoid bodies are electron dense structures found in the cytoplasm of many cells that fit the classical definition of a neoblast.

While two types of neoblast-like “stem” cells were noted, these most likely represent neoblasts in different stages of the cell cycle because they were separated based on DNA content. At the moment, general cell morphology and fluorescent dyes combined with □-irradiation are the only available parameters to sort cells, which results in the sorting of clearly mixed cell populations. The promising potential of FACS analysis awaits the identification of markers that can be used to separate cells based on molecular criteria. The best example to date of a lineage-restricted neoblast fate is the presumptive primordial germ cells. The germline stem cells derive from somatic cells because head fragments devoid of germ cells regenerate them and can eventually make oocytes and sperm.

Germ cells represent a lineage-restricted stem cell type and are indistinguishable from neoblasts at the ultrastructural level. However, these cells can be identified by their distribution and their specific expression of the planarian*nanos* homolog. *Nanos* encodes an RNA binding protein with a known role in germ cell differentiation and maintenance, and in planarians the undifferentiated *nanos*-expressing cells are located near the testes and ovaries. Importantly, silencing *Smed-nanos* by RNAi does not affect neoblast function or planarian regeneration, but abolishes the formation, regeneration, and maintenance of gonads in sexual planarians.

In addition, the germline represents the only clear case of cycling cells besides neoblasts because these cells: 1) are sensitive to irradiation incorporate BrdU ; 3) express PCNA ; and 4) stain positive for H3P in the spermatocyte cysts. Interestingly, expression of the *nanos*homolog in asexual animals revealed that although these animals do not form functional gonads, they still specify germ cells that do not divide.

The vast potential of modern molecular tools to dissect the nature of planarian stem cells. Comparing microarray expression profiles of non-irradiated and irradiated animals at either 24 hrs or 7 days post-irradiation generated a list of genes enriched in neoblasts and their progeny. These comparisons were crucial because neoblasts, and therefore neoblast-specific genes, disappear by 24 hrs after irradiation. However, genes that disappear soon thereafter represent post-mitotic cell types that are lost because no neoblasts remain to replenish them.

Three categories of neoblast-related genes were identified based on their ordered rate of disappearance following irradiation, and on their expression pattern in intact animals. WISH, BrdU pulse-chase, and double-labeling FISH experiments showed that BrdU labeled cells, which are neoblasts at the time of labeling, exhibit a stereotyped progression of cell differentiation. At early time points, 99.3±0.5% of BrdU+ cells are positive for Category 1 markers. By 2 days after a BrdU pulse, BrdU+ cells instead express category 2 markers and by 4 days they express category 3 markers. While cells expressing category 2 and 3 markers are descendents of neoblasts, the precise relationship between these cells remains to be determined.

This study unambiguously identified markers to functionally investigate the self-renewal of neoblasts and the differentiation of their progeny. However, this is just the tip of the iceberg. While the WISH pattern of each category 1 marker is similar and marks neoblasts, careful double FISH of all category 1 markers may reveal long sought-after molecular heterogeneity among neoblasts. These studies significantly expand on earlier expression analysis of irradiated animals from a related planaria species. However, the genes screened by microarray still represent only a small fraction of the planarian genome and additional early lineages remain to be discovered. Moreover, the experiments addressing the temporal regulation of gene expression were performed on intact animals.

During the homeostatic conditions studied, all differentiated adult cell types were present. How the animal senses which body parts to replace after amputation, how it selects the appropriate lineages, and how this corresponds to the lineage relationships observed in intact animals are fascinating questions for future investigation.

Gene function and neoblasts

Several neoblast studies have focused on planarian genes that encode homologs of RNA binding proteins associated with germ granules in other animals. While some planarian RNA binding proteins are expressed in a neoblast-like pattern (*piwi* homologs), some are expressed in both neoblasts and differentiated tissues, and others are expressed exclusively in the germline or in differentiated tissues. RNAi-mediated silencing of *pumilio, piwi, or bruno* homologs leads to regenerative failure, but through different mechanisms. While neoblasts are eventually lost in all 3 cases, careful evaluation of early phenotypic stages has led to important insights.

When a *bruno* homolog (*Smed-bruno-like*) is silenced, stem cell maintenance is defective while differentiation is normal. These animals can initiate regeneration and begin forming new tissue, but as neoblasts are depleted, this tissue regresses and the animals die. On the other hand, when a *piwi* homolog (*smedwi-2*) is silenced by RNAi, neoblasts proliferate, their progeny migrate, but differentiation is defective. These studies illustrate the potential to understand various aspects of stem cell control during tissue homeostasis and regeneration. They also illustrate that it is crucial to distinguish primary from secondary regeneration phenotypes, because the loss of neoblasts is a common secondary consequence of earlier defects in distinct processes.

Genes that do not encode RNA binding proteins can also be silenced to elicit irradiation-like stem-cell-defective phenotypes. After silencing *Smed"cdc23*, neoblasts arrest in anaphase, cannot divide to replace tissue during homeostasis, and the animals curl and lyse as if they had been irradiated. In addition, RNAi of a planarian innexin homolog (*smed-inx11*) abolishes regenerative capacity and leads to ventral curling. This gene is expressed in post-mitotic cells anterior to the photoreceptors much like the category 2 genes, smedinx-11 expression may reflect a transition state from neoblasts to differentiating progeny, but this supposition has not been rigorously tested.

Fate choice

In 1904, Thomas Hunt Morgan observed that anterior and posterior facing amputations both have the potential to become either a head or a tail. Scientists had to wait over 100 years for molecular understanding of how planarians sense which structure to replace.

This new insight stems from silencing the intracellular core components of the canonical Wnt signaling pathway. RNAi silencing of a planarian *□-catenin* homolog or *dishevelled* homologs induces the regeneration of a head even after tail amputation. Conversely, silencing the *adenomatous polyposis coli (APC)* homolog, which encodes an antagonist of □-catenin, causes increased □-catenin activity and consequently, a tail regenerates even if the head is removed.

These data indicate that under normal circumstances, □-catenin is a molecular switch: activity is inhibited or never initiated at anterior wounds to induce head formation and is highly activated at posterior wounds to induce tail formation. The control of □-catenin activity is also crucial for homeostasis because RNAi of *□-catenin* (*Smed-□catenin-1*) in intact animals causes neoblast progeny throughout the animal to adopt an anterior fate, leading to the transformation of other tissue types into heads. Thus, the silencing of specific planarian genes has uncoupled fate decisions of new neoblast progeny from their location in the animal.

The control of □-catenin activity does not actually specify head or tail fate, but instead anterior or posterior fate, which consequently leads to head or tail formation. This is supported by at least 3 lines of evidence. First, after cutting *Smed-□catenin-1(RNAi)* animals just anterior to the photoreceptors, they do not regenerate a new head anterior to the old one, but instead specify the anterior tissue that needs to be replaced. Second, tail fragments of untreated animals, which must regenerate a trunk and a head, specify the anterior margin of the regenerating animal before an actual head or trunk forms. Regeneration and tissue remodeling then replace the missing regions.

Third, after lateral amputations of RNAi treated animals, nearly all of the cells along the entire amputation plane adopt an anterior or posterior fate. Hence, while □-catenin activity must be kept low in the anterior and high in the posterior during lateral regeneration, this activity must be maintained at intermediate levels at intermediate positions so that neoblast progeny can adopt fates other than anterior or posterior. This implies that there may be a gradient of □-catenin activity along the AP axis or perhaps a third state (head, body, tail) of the □-catenin switch. The mechanism by which this gradient is established or this switch controlled during regeneration remains the focus of current research.

The discovery that □-catenin activity is dynamically controlled during planarian homeostasis and regeneration has provided a foothold to study how planarians recognise that the anterior or posterior structures have been removed.

Two main questions remain unresolved. First, which cells turn up or down □-catenin activity to determine head or tail identity? Is it the stem cells, their progeny, or both? Second, what are the upstream signals that control □-catenin activity? Secreted Wnt ligands and antagonists are the most likely candidates and their RNAs are expressed in a complex AP gradient, but silencing these genes has not yet led to head or tail misspecification defects. If Wnts are used to control the □-catenin switch, then what mechanisms are in place to control Wnt expression?

HEALING OF TISSUE DAMAGED BY INFLAMMATION

After inflammation has damaged tissue (when combatting bacterial infection for example) and pro-inflammatory eicosanoids have completed their function, healing proceeds in 4 phases.

RECALL PHASE

In the recall phase the adrenal glands increase production of cortisol which shuts down eicosanoid production and inflammation.

Resolution phase

In the Resolution phase, pathogens and damaged tissue are removed by macrophages (white blood cells). Red blood cells are also removed from the damaged tissue by macrophages. Failure to remove all of the damaged cells and pathogens may retrigger inflammation.

Regeneration phase

In the Regeneration phase, blood vessels are repaired and new cells form in the damaged site similar to the cells that were damaged and removed. Some cells such as neurons and muscle cells (especially in the heart) are slow to recover.

Repair phase

In the Repair phase, new tissue is generated which requires a balance of anti-inflammatory and pro-inflammatory eicosanoids. Anti-inflammatory eicosanoids include lipoxins, epi-lipoxins, and resolvins, which cause release of growth hormones.

STEM CELLS IN ANIMAL MODELS OF REGENERATION

Regeneration is arguably among the most awe inspiring biological phenomena known to exist. The history of the Western canon is populated by many examples of the indiscriminate, powerful grip regeneration has exerted on the human mind. For instance, when Lazzaro Spallanzani reported in 1768 that decapitated snails regenerate their heads, scientists, philosophers and the public alike scoured their gardens in an attempt to replicate this fascinating experiment.

It was also discovered that salamanders can regenerate limbs and tails (including the spinal cord), while planarians can regenerate entire animals from small body fragments.

Despite the longstanding interest in this biological problem, and the knowledge that animals from all walks of life perform regenerative feats, we are still in the early stages of describing these events in cellular, molecular, and mechanistic terms. However, the genetic and molecular tools to address the problem of regeneration are rapidly improving. Aside from the curiosity it normally elicits, the study and understanding of regeneration could dramatically impact the practice of medicine.

Just as relevant is the understanding derived from the investigation of **stem cells**, undifferentiated cells that have the capacity to replace themselves indefinitely and to produce specialised cell types. While embryonic stem cells divide and ultimately give rise to all the differentiated cell types of the body, adult stem cells from specific tissues are normally lineage restricted to a specific set of cell types. In order for an adult animal to replace missing structures with an exact copy of what is missing, it is clear that developmental programmes must be redeployed. However, the dynamics of cell communication and proliferation are vastly different, as are the cell types involved.

To accomplish regeneration, adult animals may invoke the proliferation of differentiated cells, the activation of reserve stem cells, the formation of new stem cells with limited capacity to self renew (**progenitor** cells), or a combination of these strategies.

Which cells in an adult animal divide and differentiate to replace the multiple cell types required during a regenerative response? While this is a very basic, indeed, fundamental question that has been formulated and reformulated through successive generations of biologists, its resiliency against experimental attacks has proved surprising, and in many cases quite frustrating. Nonetheless, it is apparent that different tissues (both within the same organism as well as the same tissues from different organisms) use different strategies to achieve tissue repair or regeneration. For example, the vertebrate liver invokes compensatory regeneration after the removal of two lobes, whereby the remaining lobe proliferates to reacquire the original tissue mass without replacing the missing lobes. In fact, regeneration can be compensatory (liver), tissue-specific (heart, skeletal muscle, liver, pancreas, lens, retina), or it can rebuild complex structures containing multiple tissue and organ types (e.g., limbs, fins, tails).

The goal of researchers studying model organisms of regeneration is to discover how these animals naturally accomplish the seemingly impossible task of restoring body parts lost to trauma. This chapter will focus on what we have learned about stem/progenitor cell identity and function from the most common, non-mammalian model organisms that invoke cell proliferation to drive regeneration. Although we will limit our discussion to bilaterian animals, we predict that significant overlap in regeneration mechanisms and concepts will emerge from the simultaneous study of regeneration and stem cells in pre-bilaterian animals such as hydra. These particular aspects of hydra stem cells and regeneration, have been exhausitively reviewed elsewhere. Still, a set of common, fundamental questions remain unanswered for all of animal regenerartion:

Where do regenerative stem/progenitor cells come from and what can they do? Given that no single animal is a model for all biological contexts, it is essential to study the multiple ways nature has solved the problem of regeneration. Therefore, the best way forward is to integrate the information derived from multiple model systems of regeneration as they are subjected to genetic, cellular, and molecular interrogation.

REGENERATION IN AMPHIBIANS: URODELES (SALAMANDERS, NEWTS, AXOLOTLS)

Background

Among the vertebrates, urodele amphibians are unmatched in their regenerative capacities. When injured, these animals regenerate an impressive array of body parts, including the upper and lower jaw, lens, retina, limb, tail, spinal cord, and intestine.

In some cases, the restoration of complex anatomy involves the formation of a **blastema**, a mass of morphologically undifferentiated, proliferating progenitor cells that is covered by epithelium and differentiates to replace the missing structures. Upon amputation, epithelial cells migrate to cover the wound, forming a **wound epithelium (WE)**. The WE thickens via distant epithelial proliferation and continued migration. The thickened structure is called an **apical ectodermal cap (AEC)**, which is thought to be similar in function to the apical ectodermal ridge (AER) that forms in the limb bud during embryonic development. While histolysis and fragmentation occurs near the wound, undifferentiated cells accumulate to form a blastema via proliferation and migration from the stump tissue. This is followed by cell specification and patterning.

But, where do blastema cells come from? Are all cells of the blastema undifferentiated and equally potent or are they restricted to become only the tissue type from whence they came? Could blastema cells have multiple, intermingling developmental origins and outcomes? Early histological studies provided evidence that cells remaining after injury "dedifferentiate" and migrate to form the blastema. Among these early studies, elegant cell tracking experiments using tritiated thymidine showed that muscle, neural sheath, periosteum, and loose connective tissue cells up to 1 mm away from the amputation plane proliferate and migrate to give rise to the blastema. In contrast, epithelial or blood cells do not appear to contribute to the growing mass of mesenchymal tissue.

The term "dedifferentiation" is often employed in the regeneration literature to describe the loss of differentiated characters of cells after amputation and their concurrent acquisition of an undifferentiated morphology during blastema formation.

In Hay's classic paper, she describes the dedifferentiation process as "the release of living cells from the confines of their previous organisation with accompanying active mitosis of these cells." However, "dedifferentiation" is easily misinterpreted to imply that a cell has attained a multi-potent undifferentiated state. Because there is currently insufficient evidence to either suggest or rule out a major role for reserve stem/progenitor cells in urodele blastema formation, the term dedifferentiation as originally coined may therefore refer to a reversal of the differentiated state, an activation of reserve stem/progenitor cells, or a combination of both.

We emphasise that the true makeup of the urodele blastema remains unknown. The extent of multi-potentiality and/or heterogeneity is presently questionable because long-term cell lineage tracing has been extremely difficult to perform in these organisms. Although blastemal cells have lost both differentiated morphology and the expression of genes associated with their tissue-specific differentiated fate (*e.g.*, muscle myosin heavy chain), it is quite possible that these cells have entered a migratory or proliferative state without changing their respective tissue identity or potentiality. In addition, recent evidence suggests that a satellite-like cell population in urodele muscle contributes to the blastema and eventually to cartilage and epidermis. Evidence of pre-existing progenitor or stem cell populations for other tissues has not been forthcoming, and thus needs to be adequately explored with modern techniques. The largest known contributor to the urodele blastema is dermal tissue, including connective tissue fibroblasts. Yet, very little is known about the heterogeneity or potentiality of this cell population on a cellular or molecular level. The recent identification of a potential progenitor cell population in the regenerating zebrafish heart illustrates the targeted type of approach that may be necessary to detect elusive blastemal progenitor cells.

In vitro studies

Recent work has examined the contribution of muscle to the blastema because myofibers/myotubes are readily identified and clearly represent a differentiated cell type. Muscle cell culture experiments are focused around Newt A1 and mouse C2C12 cells, both of which are mono-nucleated cells that can be induced to fuse and form myotubes in culture upon serum deprivation. These experiments uncovered a key difference between newt and mammalian *in vitro* derived myotubes. Newt A1 myotubes will enter S phase when exposed to high levels of serum or lower levels of thrombin-treated serum, while mouse C2C12 cells do not respond indicating the presence of an active signal in animal serum that leads to proliferation of newt A1 myotube nuclei. However, it should be noted that the extent of differentiation and, therefore, relative developmental starting point for the two cell types is not clear. For example, newt A1 cells do not exhibit striation or peripheral alignment of nuclei. In addition, *in vivo* urodele amphibian muscles contain satellite-like cells that express Pax7 and are separated from the myofiber by a basement membrane. It is doubtful that the cultured A1 myotubes contain the satellite-like cells that are present *in vivo*. Therefore, whether or not *in vitro* generated myotubes represent *in vivo* myotube behaviour needs to be fully determined. This is important because results from these cell lines have been interpreted to represent the urodele and mammalian condition. Because the cultured cells were serum starved to trigger myotube formation in the first place, an alternative interpretation of these experiments is that newt A1 cells are more flexible or "plastic" than mouse C2C12 cells. This may turn out to be the general case for urodele cells relative to mammalian cells, or it may be a feature unique to newt A1 cells, which may not be fully differentiated. In addition, while newt A1 cells do enter S phase after serum treatment, they do not go on to divide and the myotubes do not fragment.

Forced expression of the *Msx1* gene (a homeodomain protein with known repressor functions) in C2C12 myotubes causes a small fraction (5%) of cells to fragment into proliferating mononuclear cells. Under the proper culture conditions, these cells can differentiate into adipocytes (fat), chondrocytes (cartilage), myocytes (muscle), or osteocytes (bone). Because C2C12 cells are multipotent to begin with, these results should be treated with caution. Nevertheless, this was a key demonstration that mammalian myofibers can be induced to reverse their differentiated state.

Complementary studies were also carried out in newts in which primary larval limb myofibers that normally fragment upon dissociation were inhibited from doing so via *Msx1* knockdown. The data argue that *Msx1* may be necessary and sufficient for differentiated muscle to fragment into proliferating mononuclear cells. To the contrary, *in vivo* morpholino-mediated knockdown of *Msx1* in individual larval axolotl tail muscle cells had no negative effect on the ability of these cells to fragment. This discrepancy has several potential explanations, including but not limited to the differences likely to exist between *in vitro* and *in vivo* conditions, and that the muscle cells of the limb and tail could exhibit a differential requirement for *Msx1* expression during fragmentation.

To track dedifferentiation events, cultured cells can be labeled and implanted under the skin of regenerating limbs. *In vitro* differentiated myotubes, labeled with dye or viral insertion, remain stable in culture, but about 25% fragment upon implantation to generate proliferating mononuclear cells that contribute to the blastema. A few cells were eventually observed to form cartilage cells, suggesting a change in cell fate, but this was an extremely rare event. On the other hand, reserve satellite-like cells also appear to contribute to the blastema. Proliferating cells derived from satellite-like precursors were isolated in culture from adult newt myofiber explants. When tagged and implanted into regenerating adult limbs, these cells contributed to the blastema and many appeared to switch lineage into cartilage and even epidermal cells. To the contrary, another group found that implanted primary myofibers from juvenile axolotls can fragment and proliferate in the absence of satellite cells.

It is possible that these disparate results can be explained by the difference in species and life cycle stages (adult versus juvenile), or by the different criteria used to assess whether satellite cells were present. It also remains possible that both myofiber fragmentation and satellite cell proliferation contribute to the blastema *in vivo* and their relative contribution may be age and/or species dependent. A definitive explanation for these discrepancies will be important and most likely awaits *in vivo* cell tracking experiments.

In another set of implantation experiments, cells isolated from the newt heart (cardiomyocytes, CMs) were isolated, tagged, and implanted into un-amputated or amputated limbs. While CMs implanted into un-amputated limbs were stable and exhibited no special behaviour, they were activated when implanted into day 5 regenerating limb blastemas and 65% gave rise to skeletal myotubes while a few expressed a cartilage cell marker. This is clear evidence of the plasticity of the differentiated state in adult urodele amphibians, but the magnitude of fate change is unclear because the transplanted CMs continued to express desmin, a marker found in many muscle cell types. While these experiments do not rule out a role for reserve stem/progenitor cells, they clearly illustrate that a large fraction of isolated newt CMs can at least switch muscle types.

***In vivo* studies**

The most convincing support for muscle fragmentation and cell fate switching during regeneration comes from *in vivo* cell tracking experiments. Unfortunately, technical constraints imposed by *in vivo* cell imaging have restricted these compelling investigations to larval axolotls that range from 2–5 cm in size from nose to tail. Axolotls are generally considered juveniles at around 5 cm in length and adults average roughly 23 cm long. This 5/10-fold difference in animal length between larvae and adults translates into a large difference of scale in both the limbs and tails for which regeneration programmes must be deployed.

In addition, it is currently unclear whether larval and adult animals utilize the same mechanisms, as formal evidence demonstrating the equivalence of these two biological contexts of regeneration is presently lacking. Nonetheless, the *in vivo*data provide strong evidence to support the notion that differentiated cells can in fact change their functional state. These data are represented by the following three key experiments.

Experiment (1): Multinucleated myofibers were injected *in vivo* with a cell tracking dye and monitored during regeneration. Amputations that removed 50% or more of the muscle cell led to degradation, while amputations that "clipped" the muscle cell caused it to fragment into proliferating mononuclear cells. Only 15/58 (<"25%) clipped myofibers fragmented, which may indicate either an inefficient fragmentation process or point to a heterogeneous population of myofibers, some of which mononucleate more readily than others. On the other hand, since fragmentation was observed in a small number of animals, this may be a rare event that does not play a major role in blastema formation. The authors calculate that muscle fragmentation contributes to roughly 17% of the blastema. In contrast, earlier work using triploid/diploid transplants suggested that dermal tissues contribute to roughly 43% of the blastema.

Experiment (2): To track individual neural progenitor cells (glial cells) during tail regeneration, spinal cords were electroporated immediately after tail amputation to force expression of GFP under the control of a glial-specific promoter. While most of the cells gave rise to the expected neuronal and glial cell types, in 24% of the animals spinal cord cells migrated out of the regenerating spinal cord, contributed to the blastema, and gave rise to muscle. In 12% of animals glial cells gave rise to cartilage. These findings are significant because they clearly demonstrate that at least in the larval axolotl, neural progenitor cells of ectodermal origin can switch fate into mesodermally derived tissues during a regenerative response.

Experiment (3): A series of transplants and single cell electroporations were recently performed to trace the lineage of spinal cord cells during tail regeneration. Spinal cord tissue transplants from GFP(+) to GFP(-) animals showed that the cellular precursors used to regenerate the spinal cord are recruited from within 500 m of the amputation plane. The cells close to the amputation give rise to distal spinal cord cells, while cells farther from the amputation give rise to proximal cells. Single cell GFP electroporations and embryonic GFP(+) neural plate transplants revealed that in most cases, cells retain their regional identity during regeneration such that dorsal and ventral cells each give rise to cells of the same respective position. However, in 8 of 21 electroporations and 3 of 5 transplants, cells changed dorso-ventral (DV) identities.

In addition, a fraction of ventro-lateral cells near the terminal vesicle, a temporary structure that forms at the tip of the spinal cord during regeneration, migrated out of the spinal cord. These migratory cells apparently contributed to the tail blastema and gave rise to blood vessels, Schwann cells, and occasionally to muscle and cartilage cells. The results suggest that glial cells can serve as multipotential stem/progenitors during regeneration and that the terminal vesicle may represent an accumulation of de-dedifferentiated or reserve stem/progenitor cells. However, the exact nature of the transplanted cells remains in question and may include migratory neural crest cells, providing an intriguing line of questions for future investigation. These data add a layer of complexity to the regionalisation and cellular make-up of the larval tail blastema and suggest that both lineage restricted and multipotent cells exist in the regenerating urodele spinal cord. Whether this is also true during adult urodele tail regeneration is unknown.

Unanswered questions

The future is promising, but a number of challenges lie ahead. The modern and classic urodele regeneration literature is derived from a mixture of regeneration paradigms (limb, tail, and spinal cord) and from a mixture of life stages (*e.g.*, adult newt, adult and larval axolotl), which vastly affects the scale on which regeneration takes place and the cell types that may be present prior to injury. While important progress has been made, the definitive source of regenerative cells in urodeles remains unknown and is most likely a complex mix of cell types and potentials.

The current data suggest that the blastema may be composed of both de-differentiated and reserve stem/progenitor cells, but this has not yet been rigorously elucidated. Studies aimed at establishing fundamental differences and similarities between tail and limb regeneration need to be performed. These should include long-term cell lineage analyses during limb regeneration such that issues of cell potentiality and the relative contribution of cells to the final regenerated structures can be unambiguously resolved. In addition, studies that incorporate molecular dissections of the regionalisation that occurs during proximal to distal patterning of the early limb blastema need to be expanded. Until cells are clearly marked *in vivo* and lineage traced, the nature of the limb blastema will remain enigmatic and controversial.

Likewise, much has been made of the ability of the multi-nucleated urodele myotubes to fragment and produce proliferating mono-nucleated cells that contribute to the blastema following injury. While the *in vitro* and *in vivo* data show that muscle fragmentation and proliferation occur, and that some cells switch identity during regeneration, it remains unclear as to what role this plays in the regeneration process. Because muscle only contributes to roughly 17% of the blastema, future research should focus on *in vivo* cell characterisation/tracking of other cell types, including dermal cells, to determine their contribution to the blastema and to assess their differentiation potential. Moreover, a potential role (compensatory differentiation) for lineage switching can be deduced from early experiments showing that limbs devoid of bone regenerate normal limbs, including the missing bone. This suggests that the blastema can regenerate structures absent from damaged pre-existing tissues. These experiments clearly need to be revisited with modern cell tracking techniques.

The *in vivo* and *in vitro* data alike suggest that some differentiated cells of urodele amphibians display plasticity, as they are able to convert from one lineage to another. The question remains as to how differentiated these cells are to begin with. One potential approach to assess the extent of the differentiated state is to observe the epigenetic state of the genome. Myotubes induced from newt A1 and mouse C2C12 cells appear differentiated and no longer respond to growth factors, but the epigenetic state of the genome is completely unknown for these cells. What is the state of genome methylation and other epigenetic markers of differentiation in newt A1 and mouse C2C12 cells? This same question about epigenetics can be asked of adult *vs.* larval urodele amphibians and should be asked in all regeneration contexts. Is there a difference in the epigenetic differentiated state that can account for the flexibility of cells in their response to injury? What are the epigenetic differences between the adult and larval axolotls? In fact, it remains possible that data from larval cells and tissues may not apply to the adult context. Differences are bound to exist between the limbs of a 3cm-long larval axolotl and a 25cm-long adult. For example, de-differentiation in the adult newt limb does not begin until about day 4 or 5 while the limb blastema of a larval axolotl is already subdivided into at least 3 proximal-distal zones by day 3 post-amputation. Because most if not all of the *in vivo* experiments were performed in larval animals, repeating these experiments in adult newt and axolotl limbs and/or tails will be extremely informative.

REGENERATION IN AMPHIBIANS: ANURANS (FROGS, TOADS)

The frog has been a mainstay in the field of developmental biology for many years. As a research model, this animal has provided researchers with key insights into how animals coordinate the progression from a single fertilised egg into billions of organised, communicating cells that function in the complex tissues and organ systems of an entire animal. To study frog development, important tools such as transgenic overexpression, were developed to allow a detailed molecular interrogation of frog biology. These tools are now proving useful for the study of frog regeneration.

Anuran amphibians can regenerate limbs and tails as tadpoles. This regenerative ability rapidly declines during differentiation and metamorphosis at stage 52, such that by stage 56 differentiated cell types and ossified bones are present and the regenerative response has diminished. This illustrates a common theme of correlation between cell plasticity and regenerative ability. Tadpoles can only regenerate complex structures while they are going through a period of large-scale morphological change, including limb development and tail regression during metamorphosis. This suggests that regeneration in anuran amphibians may depend upon the presence of undifferentiated cells, which are no longer present once differentiation has set in. Does anuran "regeneration" represent the ability to regenerate, or instead the ability to forge on with development following damage ? Whether anuran limb or tail regeneration can be considered equivalent to adult tissue regeneration remains an open question.

Tadpole tail and limb regeneration

Besides the loss of regeneration at stage 56, there is a refractory period in which the regenerative ability of the tail is lost between stages 45 and 47 (4–6 days of development), but is regained after stage 48. Therefore, the anuran amphibian provides a model system that can be used to study the transition between the loss and gain of regenerative abilities. For example, the blockage of the BMP or Notch signaling pathways inhibits normal tail regeneration while overexpression rescues regeneration during the refractory period. *Msx1* is a direct target of BMP signaling and forced expression of a hyperactive form of *Msx1* is sufficient to allow tail regeneration during the refractory period. However, manipulation of these pathways did not allow regeneration of late stage tadpoles, again implicating the need for responsive cell types that are likely present prior to state 56, but absent thereafter. Regeneration of the anuran amphibian tail proceeds through the formation of an undifferentiated blastema-like structure. However, recent studies using GFP(+) tissue transplants and Cre-Lox mediated cell tracking have shown that each tissue of the frog tadpole regenerates in an independent manner, giving rise to the same tissue during regeneration.

Unlike urodele amphibians, there is currently no evidence to suggest that anuran amphibian cells cross lineage boundaries during regeneration, and unlike urodele muscle, anuran tail muscle clearly regenerates from a satellite stem cell population. Interestingly, overexpression of a constitutively active form of Notch (NICD) during the refractory period leads to the regeneration of tails with no muscle. Notch signaling is a key regulator of satellite cell fate in mammals and its role in satellite cells during tadpole tail regeneration remains to be determined. What coordinates the proliferation and differentiation of multiple tissue types, and why does regenerative ability decline despite the continued presence of satellite cells?

In anurans, as in urodeles, regeneration depends on the formation of a WE and an AEC. The formation of the WE and properly specified AEC is critically important and during periods of lost regenerative ability, amputation leads to the formation of a skin-like epithelium instead of a wound epithelium. WE and AEC specification are tightly coordinated with blastema formation using at least 4 signaling pathways. Disruption of BMP, Wnt, or Notch signaling causes the loss of regenerative abilities by affecting either the formation or maintenance of a properly stratified AEC structure and the loss of *Msx1* expression in the underlying mesenchymal tissue. These pathways appear to control the expression of coordinated FGF signals, FGF-8 and FGF-10, which are also likely to be key players in the epidermal/mesenchymal interactions that drive early blastema formation, proliferation, and regenerative outgrowth. An elegant mix of pharmaceutical and transgenic overexpression strategies recently showed that regeneration of the frog tail requires FGF and canonical Wnt signaling. These studies also suggest that Wnt signaling functions upstream of FGFs and that both pathways are inhibited when noggin, which interferes with BMP signaling, is overexpressed. While it is clear that signaling pathways play an essential early role to establish proper epidermal/ mesenchymal interactions, and that each tissue type is derived from lineage-restricted cells, it remains unclear exactly which cells of which specific tissues respond to which signals.

BASIC CONSIDERATIONS FOR *IN SITU* TISSUE REGENERATION

The success of *in situ* tissue regeneration relies on effective recruitment of host stem or progenitor cells into the implanted biomaterial scaffolds and induction of the infiltrating cells into tissue-specific cell lineages for functional tissue regeneration. To achieve this, a target-specific scaffolding system, serving as a template, needs to be designed in order to enable ('instructs') the fate of the recruited host cells to proliferate and differentiate into a desired tissue type.Sustained delivery of biological cues, such as bioactive molecules, from the implanted scaffold could play an important role in guiding host cells to form a well-integrated functional structure. Moreover, a well-designed combination of biological cues with biomaterial scaffolds would provide appropriate microenvironments for efficient cellular specification within the implanted scaffold.

HOST CELL SOURCES FOR *IN SITU* TISSUE REGENERATION

It has been demonstrated that adult stem cells that contain self-renewal and differentiation capability can be isolated from various tissues and organs, including brain, liver, circulating blood, heart, skin, kidney, muscle and fat.Most adult stem cells are quiescent and reside in a specialised microenvironment, which is called a 'stem cell niche'.

In response to regulatory signals that originate from tissue injury, these stem cells become activated and begin repairing process. In addition to tissue-specific adult stem cells that are primarily responsible for tissue regeneration processes, bone marrow-derived stem cells have been identified as important cell sources that contribute their regenerating capacity to other tissues.

The bone marrow harbors multiple distinct stem/progenitor cells that include hematopoietic stem cells (HSCs), mesenchymal stem cells (MSCs) and endothelial progenitor cells (EPCs). HSCs are responsible for the production of all circulating blood cells such as myeloid, erthyroid and lymphoid lineages. An important role of the HSC population for tissue regeneration is to provide paracrine bioactive factors to regenerative cells and occasionally transdifferentiate into desired tissue-specific lineages. Another cell population contained in the bone marrow is stromal cells or MSCs that exhibit multipotent capabilities to differentiate into a variety of cell types *in vitro* and *in vivo*.

Interestingly, many reports have shown that MSC populations that express similar set of cell surface antigens can be isolated from bone, cartilage, muscle, bone marrow stroma, tendon, fat and other connective tissues, and these MSC populations have been widely used in preclinical and clinical applications for tissue regeneration. These cells are known to modulate the immune system and/or provide trophic factors necessary for tissue-specific regeneration. However, the specific identity of these MSC populations is still unclear and further studies are required for understanding the origin and contributions of these cells to tissue regeneration. EPCs are another important cell source that are actively involved in promoting angiogenesis at the injury site. Facilitating neovascularisation through EPCs is particularly beneficial in tissue regeneration and ischemic tissue injury.

Macrophages that play an important role in inflammatory response and foreign body reaction are found at the injury site during the tissue repair and remodeling process. Recent reports have shown that macrophages are an important determinant during tissue remodeling in the context of regenerative medicine.

The proinflammatory macrophages, designated as an M1 phenotype, are involved in chronic inflammation and foreign body reactions, whereas M2 macrophage phenotypes are associated with anti-inflammation, immunomodulation and tissue remodeling processes. A better understanding of the mechanisms underlying differential infiltration of the macrophage populations into scaffolds will aid in controlling the specific type of macrophage recruitment and may result in beneficial effects on the desired tissue regeneration.

Although certain types of host cells have been identified in inflammatory responses and foreign body reactions, cell populations that infiltrate into biomaterial scaffolds are poorly understood. It is important to investigate the possibilities to use the body's biologic and environmental resources for tissue regeneration *in situ*. Recruitment of host cells into an implanted scaffold as part of tissue repairing process has been examined. In our previous study, poly(glycolic acid) nonwoven scaffold was used to address this dogma. This biomaterial implant was designed to enhance diffusion and accommodate host cell infiltrates into the highly porous structures. The results of this study showed that the number of host cells infiltrating into the implant increased for up to 3 weeks after implantation and began to decrease thereafter, as collagen accumulated to fill the pores of implanted scaffold. Interestingly, we observed that a small proportion of infiltrated host cells within the implants had multilineage potential. These results indicate that some of the host stem cells that mobilized into the biomaterial were multipotent, and given an appropriate microenvironment, they differentiated into tissue-specific cell lineages at the implant site.

PROTEIN DELIVERY SYSTEM

When bioactive molecules are administered into the injury site as a bolus injection, most of the factors tend to lose their biological activities because of enzymatic digestion in the body. To overcome this limitation and maintain effective concentrations of molecules in the local microenvironment, sustained release of bioactive cues can be accomplished by encapsulation within a scaffolding system through physical or chemical binding. Release pattern of the incorporated bioactive factors can be controlled by scaffold modifications through changing physical properties, temperature, pH and material degradability.

In particular, a scaffolding system for *in situ* tissue regeneration needs to possess an appropriate microenvironment that is able to recruit host stem and progenitor cells into the implant and support the expansion and differentiation into a desired tissue type. For this purpose, multiple factors need to be delivered to a target site because of the complexity of the microenvironment.

In one study, a multiple protein delivery system was developed for accelerating vascularisation and functional tissue formation, based on the fact that functional regeneration of tissues and organs is typically induced by the action of a number of growth factors. The investigators reported that a new polymeric system facilitated the tissue-specific delivery of two or more growth factors, and enabled sustained release of bioactive molecules with different release kinetics for effective tissue regeneration. Similarly, a recent study demonstrated various delivery methods of bioactive molecules for controlled release over time. Sustained release of multiple molecules mimics *in vivo* tissue regeneration and it contributes to effective and functional tissue regeneration.

Another study used a gelatin-based scaffold to deliver four different bioactive molecules, namely VEGF, angiopoietin-1, keratinocyte growth factor and platelet-derived growth factor-BB. The delivery of these molecules induced an increase in angiogenesis with a potential for promoting tissue regeneration.

However, the current delivery systems are limited to local delivery via release from the implanted scaffold that results in accumulation of migrated cells mostly at the periphery of the scaffold, delaying cell infiltration into the interior of the scaffold.

Consequently, the unbalanced cell localisation prevents successful tissue regeneration. It is evident that a new delivery approach that facilitates efficient recruitment of host stem cells needs to be developed.

Towards this goal, we have developed a new delivery method, in which combined systemic and local delivery of multiple factors (SP and SDF-1α) were used to enhance recruitment of host stem cells such as MSC and HSC populations from two different compartiments (circulating bloods and resident sources) into target scaffolds *in vivo*. This strategy consists of two steps: (1) to increase mobilisation of bone marrow-derived MSCs and HSCs by systemic SP injection and (2) to enhance recruitment of two different populations of the SP-stimulated bone marrow-derived cells and resident stem cells into the implanted scaffolds via local release of SDF-1α from the scaffolds. Our results showed that this combination delivery system significantly enhanced host stem cells such as $CD29^+CD45^-$ MSC-like cells, $CD146^+$a-SMA^+ pericytes and c-kit^+ cell-included HSC population into the scaffolding system, indicating the effectiveness of the combination delivery system for *in situ* tissue regeneration.

APPLICATIONS OF IN SITU TISSUE REBIRTH

The concept of *in situ* tissue regeneration has been translated into various therapeutic applications. Various biomaterial scaffolds have been used for this purpose in the form of injection or implantation. Although many technologies are at the early stage in investigations, several technologies have been successfully performed in preclinical animal models and clinical applications with satisfactory outcomes.

BONE

Osteogenic repair from bone loss has benefited from techniques of *in situ* tissue regeneration. The required properties for a bone-specific scaffold are temporal and mechanical load bearing within the tissue defects. Moreover, it should minimize immune and/or inflammatory response. Biomaterials widely used for *in situ*bone regeneration include calcium phosphate, calcium sulfate and hydroxyapatite.

As bone tissue is composed of these materials, it would be natural to consider using such materials for scaffolds for bone regeneration. This is because of their close chemical and crystal resemblance to the mineral phase of bone, demonstrating excellent biocompatibility and osteoconductivity. The commonly used bioactive molecules for bone regeneration include bone morphogenetic protein-2, transforming growth factor-β, basic fibroblast growth factor and VEGF.

In some cases, vital growth factors can be incorporated into the scaffolds to exert their osteoinductive and vascularisation properties.Biomaterials ranging from natural polymers, such as alginate, fibrin or gelatin, to synthetic polymers, such as poly(lactic acid) and PLGA, have been fabricated with either single or multiple bioactive molecules. These scaffolds have demonstrated an ability to stimulate and induce neighboring bone marrow stromal cells and enhance bone tissue formation. Many clinical applications have been conducted for bone regeneration *in situ* using calcium phosphate cements, collagen gel or sponge combined with clinically approved bone morphogenetic protein-2.

Cartilage

Injured and damaged cartilage tissue can lead to severe arthritis because of low natural healing capability when compared with other types of tissues. In early studies, cartilage tissue constructs were easily produced by seeding chondrocytes onto scaffolds. However, when engineered cartilage constructs are implanted, a serious compatibility issue such as poor integration with the host tissue is often observed. As such, a recent study showed that successful cartilage regeneration can be achieved using a cell-free scaffolding system. In this study, scaffolds consisting of biodegradable PLGA polymer were incorporated with plasma and hyaluronic acid, and implanted into microfractured cartilage tissue.Consequently, the implanted constructs facilitated the migration of bone marrow-derived stem cells that led to the formation of neo-cartilage tissue. More recently, Lee *et al.* demonstrated that an entire articular surface of the synovial joint can be regenerated without cell transplantation using three-dimensional poly(å-caprolactone) and hydroxyapatite composites fabricated by solid free-form technique. These scaffolds were incorporated with transforming growth factor-β3 and implanted into a rabbit model. Regeneration of new and avascular cartilage with vascularized subchondral bone tissue was evident. This result was shown to be effective in regenerating cartilage tissue by recruiting host stem cells to the site of the implants.

Skeletal muscle

Muscle tissue is the largest tissue mass in the body. Skeletal muscle accounts for ~45% of total body weight. Skeletal muscle tissue contains bundles of myofibers that function by contracting with motor nerve stimulation. Minor muscle injury because of exercise and weight lifting is easily restored by natural regenerative processes. However, if >20% of the muscle is lost, spontaneous recovery will not occur, leading to loss of muscle function. If the injury is not properly treated, skeletal muscle weakness and atrophy will occur.

Cell-based approaches have offered new opportunities for restoring muscle function because of severe muscular injuries. Muscle satellite cells have been identified as a cell source for muscle tissue regeneration owing to their self-renewal capabilities and muscle-specific differentiation following muscle injury. In addition to muscle satellite cells, several other stem cell populations, such as muscle-derived stem cells, pericytes, muscle-resident macrophages,EPCs and bone marrow-derived MSCs have been used for muscle tissue engineering. The roles of these cell populations are critical for efficient muscle regeneration, by promoting angiogenesis and maturing neovasculatures, secreting myogenic trophic factors and modulating inflammation for reduced fibrosis.

Several studies have been performed to regenerate muscle tissue *in situ*. In one study, an alginate gel-based dual delivery system was used to deliver insulin-like growth factor-1 and VEGF for the enhancement of functional muscle regeneration. The roles of sustained release of insulin-like growth factor-1 and VEGF from the scaffold are mobilisation and manipulation of satellite cells and inducing efficient angiogenesis for functional muscle regeneration, respectively. In another study, a collagen-based sponge scaffold was used to treat rabbit hind limb muscle injury.

At 24 weeks after implantation, the control group (without scaffold) showed poor structural regeneration with severe scar tissue formation at the site of injury, whereas the scaffold-implanted group showed mild focal adhesions and new muscle tissue formation. One important consideration for muscle regeneration *in situ* is how to create aligned and organised muscle fibers within the implanted scaffold. Alignment is critical for newly regenerated muscle fibers to exert normal physiological muscle function in response to nerve stimulation. Recently, a scaffolding system consisting of unidirectionally aligned fibers was developed by electrospinning techniques using poly(å-caprolactone) and collagen as base materials. When skeletal muscle cells were seeded on the scaffolds, cellular alignment along the polymer fibers was evident with formation of organised muscle fibers upon differentiation. This result indicates that fabrication of muscle-specific scaffolds may be important for *in situ* muscle tissue regeneration.

BIOMATERIAL SCAFFOLDS FOR *IN SITU* TISSUE REGENERATION

Creation of bioengineered tissue requires a scaffold, which provides structural support until the mobilized cells form functional tissue *in vivo*. Although the properties of scaffolds may vary depending on the targeting tissues, the general requirements of a scaffolding system are biological stability, biodegradability and temporal structural integrity. The scaffold's internal architecture should provide adequate permeability for establishing functional vascularisation following implantation.

The latter is critically important as this porous structure can not only facilitate space for the recruited cells to reside, but also permit incorporation of bioactive molecules and biophysical cues that enhance cell migration, proliferation and differentiation to produce a biofunctional host stem cell niche.To design a tissue-specific scaffolding system for *in situ* tissue regeneration, the scaffolds should possess the ability to (1) regulate inflammation for minimized fibrotic formation, (2) utilise host microenvironment for recruiting host stem/progenitor cells and (3) control tissue-specific cell differentiation within the scaffold.

Biomaterials used for scaffolding can be naturally derived or synthetic polymers. Natural materials include polysaccharides and proteins. Polysaccharides that have been widely used for this purpose include cellulose, alginate, hyaluronic acid, starch, dextran, heparin, chitin and chitosan. Proteins are the primary components of tissues or organs and have been used for various biomedical applications. Collagen, which is the most abundant protein in mammals, has been used as scaffold materials because of the ease of processing as well as its ability to induce minimal inflammatory and immune responses. Collagen has been approved by the US Food and Drug Administration (FDA) for many types of biomedical applications, including wound dressings and artificial skin. Collagen can be configured into various structures such as films, fibers and sponges. Decellularised collagen-rich tissue scaffolds have received much attention recently because of their ability to maintain microtissue architecture. In addition, these acellular tissue matrices have been shown to support cell ingrowth and tissue regeneration.

Synthetic polymeric biomaterials such as biodegradable polyesters, including poly(glycolic acid), poly(lactic acid) and poly(lactide-*co*-glycolide) (PLGA), are widely used in tissue engineering and regenerative medicine. The use of these polymers was approved by the FDA for human use in a variety of applications, including surgical sutures. These biodegradable polymers are nontoxic during the degradation process *in vivo*, and are eventually removed from the body in the form of carbon dioxide and water. These polymers possess thermoplastic properties and can be easily fabricated into various configurations with controlled microstructure and porosity using a number of processing techniques, including molding, extrusion, solvent casting, phase separation techniques and gas foaming techniques. More recently, electrospinning techniques have been developed to quickly create highly porous scaffolds in various conformations, including nanostructures. Other biodegradable synthetic polymers used for tissue regeneration applications include poly(anhydrides) and poly(ortho-esters).

Biochemical signaling for *in situ* tissue regeneration

A key to *in situ* tissue regeneration in the initial stage is proficient recruitment of host stem or progenitor cells into an implanted scaffold. However, adult stem cell populations in the body are generally too low in number to have a significant impact on acceleated tissue regeneration. In most cases of tissue regeneration, bone marrow-derived stem cells have contributed to regeneration, and therefore it is worthwhile to target these cells to be effectively mobilised into the peripheral blood system. During this mobilisation, homing and engraftment process, other cytokines and chemoattractants are able to increase the efficacy of migration to the injury site.

Substance P (SP) is a neuropeptide that functions as a neurotransmitter and neuromodulator. A recent report showed strong evidence that released SP after corneal injury was able to mobilise a high number of stromal-like $CD29^+$ cells (MSC-like cells) from bone marrow into peripheral blood and drive migration and participation in the repair processes of the cornea injury. This study shows that under the optimal culture conditions *in vitro*, the SP-induced $CD29^+$ MSC-like cells are able to demonstrate multidifferentiation capability by forming bone, cartilage and fat cells. These results indicate that SP is expected to play a positive role in tissue repair based on the multipotency features. In terms of therapeutic aspects, the use of SP appears to be a cost-effective treatment because of high efficacy of host MSC-like cell mobilisation with a single injection.

A recent report also stressed positive roles of SP in reparative neovascularisation. In this report, patients with myocardial infarction showed high concentration of SP in the blood, which increased host progenitor cell mobilisation, whereas suppression of SP levels resulted in a decreased number of host therapeutic progenitor cells. The study team concluded that SP-based nociceptive signaling may represent a possible target of regenerative medicine. Therefore, the use of stem cell-stimulating factors such as SP is a possible approach to accelerate the neovascularisation process.

It is well known that retention of stem or progenitor cells in the bone marrow is through the interaction between the C-X-C chemokine receptor 4 (CXCR4) on the surface of stem/progenitor cells and stromal cell-derived factor-1α (SDF-1α) on the surface of bone marrow stromal cells. When the retention axis is disrupted, the progenitor cells are released from the bone marrow stroma and mobilized into the peripheral blood. One of the mobilisation-accelerating factors is granulocyte-colony-stimulating factor (G-CSF) and it has been widely used for clinical trials.It was reported that $CD34^+$ HSCs can be effectively mobilized into the peripheral blood from the bone marrow through disruption of the SDF-1/CXCR4 axis.

In an approach for *in situ* tissue regeneration, several reports showed that G-CSF has been used as a single injection or directly incorporated into the implanted scaffold. A single injection of G-CSF for tissue regeneration has been conducted to accelerate EPC recruitment on implanted small-diameter vascular constructs.It was shown that administration of G-CSF induced significant recruitment of $CD34^+$, $CD133^+$ EPCs into the vascular graft, generated endothelium and inhibited neointimal hyperplasia of a small-diameter heparinized decellularized vascular graft. A recent approach showed the efficiency of released G-CSF from a hydrogel scaffold for enhancing EPC mobilisation.

This group developed a hydrogel system incorporating G-CSF and showed that intramuscularly injected hydrogel significantly enhanced mobilisation of $CD34^+CD31^+$ EPCs into the blood, as compared with a G-CSF bolus injection or hydrogel injection only. In addition, AMD3100, an antagonist of CXCR4, has been used singly or in combination with G-CSF to enhance mobilisation of HSCs and progenitor cells.

Another report describes the use of AMD injection for the treatment of myocardial infarction. AMD treatment enhanced mobilisation and recruitment of EPCs to the neovasculature. In addition, a combined treatment of G-CSF and AMD has resulted in efficient mobilisation of monocytes and stimulation of angiogenesis at ischemic sites.

Stem cell factor is an endogenous ligand for the tyrosine kinase receptor c-*kit*, which is expressed on HSCs. Recombinant stem cell factor has been shown to act in synergy with G-CSF in mobilisation of bone marrow-derived

HSCs. Regulation of selective mobilisation of different populations of stem cells has been tested.This study showed that treatment with CXCR4 antagonist (AMD3100) effectively mobilizes HSCs, but not EPCs or stromal cells. However, pretreatment with vascular endothelial growth factor (VEGF) resulted in EPC and stromal cell mobilisation, whereas HSC mobilisation was reduced. These results suggest that multiple intersecting signaling pathways regulate the proliferation and mobilisation of bone marrow-derived stem cells for efficacious tissue regeneration.

Direct targeting of the stem cell niche is another approach to induce stem cell mobilisation for promoting tissue regeneration. In bone marrow, one component of the HSC niche is osteoblasts. It was shown that stimulation of the parathyroid hormone receptor promotes osteoblast proliferation and secretion of paracrine factors that, in turn, resulted in an increase in the number of HSCs. These studies indicate that direct targeting of osteoblasts can modify the activity of HSCs in the bone marrow.

In addition to stem cell-stimulating factors that mobilize host stem/progenitor cells in the body, bioactive molecules that induce engraftment of the mobilized host stem cells into desired tissues or organs for repair are considered as important cues for efficient *in situ* tissue regeneration. One representative chemoattractant is SDF-1α that has been shown to attract MSCs and HSCs to injured tissues through CXCR4 (SDF-1 receptor) expression. In peripheral blood where the expression of CXCR4 is at a low level, a pool of both MSCs and HSCs can be maintained at a balanced level in distant parts of the body. However, tissue damage or other mobilisation cues could mobilize these cells to peripheral blood and to the injury site.

Thus, it is possible that sustained release of chemoattractants such as SDF-1α contained within an implanted scaffold could generate a high concentration gradient of these factors and drive efficient stem cell migration into the implant. For example, a previous report showed that local release of SDF-1α was observed from heparinized collagen sponge-enhanced recruitment of HSCs into subcutaneously implanted scaffolds. One recent study demonstrated an interesting approach to utilize the effects of SDF-1α on recruitment of host stem cell into implanted scaffolds. This group incorporated SDF-1α into a biodegradable PLGA scaffold and implanted subcutaneously in mice. They showed that local release of SDF-1α induced efficient recruitment of host stem cells (SSEA-4^+ cells) in the implanted scaffold. In addition, the scaffolding system resulted in a reduced inflammatory response. This approach has also been applied to brain injury for the recruitment of neural progenitor cells. For accelerated brain regeneration from cavitary brain lesions that fail to recruit endogenous neural progenitor cells, the authors developed an injectable scaffolding system that consists of gelatin-hydroxyphenylpropionic acid hydrogels and dextran sulfate/chitosan polyelectrolyte complex nanoparticles to deliver SDF-1α to the cavitary brain lesion region. They demonstrated the initial feasibility *in vitro* by showing that release of SDF-1α from the incorporated gel scaffold system significantly enhanced infiltration of neural progenitor cells, when compared with hydrogel only or vehicle controls, indicating that this scaffolding system is a promising approach for neural tissue repair.

SDF-1α is known to be easily damaged by matrix metalloproteinase-2. To address this issue, SDF-1α molecule was engineered to be resistant to protease activity and the engineered SDF-1α was tethered to self-assembling peptides to form nanofibers. In this study the authors showed that the injected nanofibers containing protease-resistant SDF-1α molecules enhanced recruited EPCs, increased capillary density and functionally improved cardiac function. In an attempt to protect SDF-1α effects from endogenous inhibitors, an alternative approach showed that targeting of SDF-1α inhibitor improved SDF-1αeffectiveness in stem cell engraftment into the damaged heart. The dipeptidyl peptidase IV is known to cleave SDF-1α, therefore inhibition of dipeptidyl peptidase IV by the small-molecule diprotin

A increases the concentration of SDF-1α in the heart following myocardial infarction. This resulted in increased progenitor cell recruitment to the ischemic myocardium and improved neovascularisation and ventricular function. Targeting of these inhibitors represents an effective strategy for regenerative medicine.

Monocyte chemotactic proteins (MCPs) are known to direct bone marrow-derived stem cells into the injury sites. One study shows that MCP-3 recruits MSCs into myocardial infarction site through activation of CC chemokine receptors. In addition, MCP-1 and MCP-5 have been reported to be CCR2-activating chemokines that recruit bone marrow-derived macrophages into toxin-induced muscle injury site to restore angiogenesis and muscle regeneration. A neuropeptide, galanin, has also been identified to play an important role in facilitating bone marrow-derived MSC migration through activation of galanin receptor.

EXTRACELLULAR MATRIX AND SCAFFOLD

A matrix or culture is needed to facilitate cell growth. *In-vivo*cells are arranged in the extracellular matrix (ECM) compromising a complex structural entity surrounding and supporting cells.The ECM is necessary in order for certain cells to carry out their functions. The ECM has 3 major components: structural proteins (mainly collagen); proteoglycans and hyaluronan; and specialised multiadhesive proteins. Every tissue contains itsown type of ECM specialised for its particular function. The amount of specific components varies according to the function of the tissue.As a result of the minimal availability of extracellular space in the epidermis, cells are tightly bound to the basal lamina, which is a thin matrix overlying the loose connective tissue in the dermis.

Proteoglycans are macromolecules consisting of a core protein attached to several polysaccharides (glycosaminoglycans). Proteoglycans are able to bind cells to the matrix and bind growth factors (eg, fibroblast growth factor [FGF], transforming growth factor beta [TGF-β] to glycosaminoglycans, thus preventing extracellular protease degradation of growth factors.

Organisation of all ECM components depends on the presence of binding proteins in the ECM. Fibronectins are another class of matrix proteins that can bind to integrins and play an important role in attaching cells to the ECM.

In tissue engineering, much research focuses on developing artificial matrices or scaffolds. Scaffolds can be permanent or temporary depending on the application. Permanent scaffolds are not biodegradable and remain incorporated in the body. These scaffolds are necessary in situations where continuous strength is needed to allow tissue to retain its shape. Today, mostly temporary and biodegradable scaffolds are used ultimately leaving only the engineered tissue *in situ*.

Research has shown that combining collagen and hyaluronic acids in scaffolds, in addition to using hyaluronic acid matrices, has been successful in tissueengineering of cartilage and dermal tissue regeneration.

BIOREACTOR

A bioreactor can be defined as any apparatus that attempts to mimic and reproduce physiological conditions in order to maintain and encourage cell culture for tissue regeneration. Cell-culture parameters such as temperature, pH, biochemical gradients, and mechanical stresses should be continuously controlled during the maturation period.

Bioreactors have already improved the processing and the final results of skin regeneration. It is essential that bioreactors are designed and fabricated following specifications that differ from tissue to tissue.

Cytokines

Cytokines are polypeptides that can bind to cell-specific receptors present in cell membranes of target cells triggering cell-specific responses. These polypeptides usually exist as inactive precursors that need to be cleaved and bound to the ECM to become active. By binding receptors cytokines can exert a signal transduction cascade via second messengers leading to changes in gene expression mediating specific cellular responses. Responses of binding a cytokine can include altering the expression of membrane proteins, secretion of effector molecules, and proliferation. Cytokines are not cell type specific—a single cytokine can act on multiple cell types and different cell types can secrete the same cytokine, despite differing responses between different cell types.

Epidermal growth factor (EGF), fibroblast growth factor (FGF), platelet-derived growth factor (PDGF), and TGF-β are the 4 most important growth factors in wound healing. EGF stimulates growth and migration of keratinocytes *in vitro* and epidermal regeneration *in vivo*. Epidermal growth factor has already been applied to delivery systems leading to the enhancement of epidermal regeneration in partial-thickness wounds and second- degree burns. Once applied to a hydrogel, significantly faster wound healing was observed.

Fibroblast growth factor (FGF) acts on fibroblasts. Of the 20 FGF family members,FGF-β is especially known as a potent stimulator of endothelial cells that induces angiogenesis.

Platelet-derived growth factor (PDGF) is also important in wound healing as it attracts various cells to the lesion, including fibroblasts, smooth muscle cells, neutrophils, and macrophages. It also stimulates secretion of growth factors by macrophages and matrix products, such as fibronectin, hyaluronan, collagens, and proteoglycans. PDGF plays an important role during the remodeling phase by stimulating the fibroblast production of collagenases. Recombinant PDGF is therapeutically applied to chronic diabetic ulcers enhancing wound healing.

The transforming growth factor-β (TGF-β) super family consists of more than 40 members with a similar structure. Almost all cells in the human body secrete inactive TGF-β. In order to bind its receptor,TGF-β must be activated. Binding to its receptor initiates a cascade mediated by activation of a protein kinase.Research has shown that incorporation of TGF-β in a collagen scaffold resulted in a faster epithelialisation and contraction rates in full-thickness skin defects in rabbits. Research focuses on the development of delivery systems for TGF-β that can gradually release TGF-β.

TGF-β1 is a pleiotropic growth factor, which has reguvanlatory effects on many different cell types. For instance, it plays an important role in cell proliferation and differentiation, bone formation, angiogenesis, neuroprotection, and wound repair. However, long exposure to high doses of TGF-β1 results in fibroses and hypertrophic scars.

Literature regarding cytokine manipulation of proliferative scars has shown that TGF-β2 may be involved in the development of tissue fibrosis.

In wound healing, TGF-β3 has been demonstrated to attenuate type 1 collagen synthesis and reduce scar tissue formation. Despite previous meritorious efforts to investigate the therapeutic potential of TGF-β3, its effective use is limited by a number of common shortcomings, such as a short half-life, *in-vivo* instability, and relatively inaccuracy of the delivery system.

WOUND HEALING AND TISSUE ENGINEERING

History

When considering wound healing historically, a trend can be seen throughout the centuries representing a shift from dry wound healing to wet wound healing. The oldest medical texts describing ancient medicine are written on papyrus that date from 1500 to 1000 BC and are based on much older initial documents dating from approximately 3000 BC. These documents indicate that in 3000 BC hemorrhages were cauterized to stop the bleeding. Around 2000 BC bandages began being used for approximating the wound edges. Later, around 1000 BC, Homer described 147 wounds in the Iliad.

Throughout history different materials have been used for bandaging wounds. The indication of bandages shifted as Celsus in 25 AD described signs and symptoms of inflammation. Since then, bandages were used for prevention of the described symptoms and signs. In the Middle Ages, Paracelsus wrote of dressing wounds made with cotton and gauze. In the 19th century, Pasteur favored the use of dry dressings to help prevent infection.

In 1962,Winter demonstrated that partial-thickness wounds reepithelialized more rapidly under occlusive dressings because occlusive dressings maintained a moist wound surface.This landmark study showed that a moist environment accelerated the re-epithelialisation process. Numerous studies followed that demonstrated wound occlusion and moisture increased all phases of healing. Occlusive dressings are designed to keep the wound bed moist with exudate. The basic concept behind moist wound healing is that the presence of exudate in a wound will provide an environment that stimulates healing by delivering a range of cells and cytokines necessary for wound repair. With these findings, use of dry dressings became less popular, and lead to the development of occlusive dressings such as, hydrocolloids, hydrogels, semipermeable films, alginates, and foams.

Since 1992, research indicates that wet wound healing might also accelerate the healing process when compared to dry wound healing. Later, in 1995 it was shown that wet wound healing showed less subepidermal inflammatory cells when compared to moist wound healing represented by a hydrocolloid dressing.

Wet wound healing uses a plastic transparent chamber filled with a medium that is attached to the wound.The chamber makes sure that the wound is constantly in contact with the medium and allows visible sight of the wound bed. Fluid chambers function as a quasi mini-incubator that make it possible to inject antibiotics and analgesics to the medium. With developments in tissue engineering emphasising the importance of adding growth factors, wet wound healing is suitable for allowing cytokines and cells to be added to the medium.

Tissue-engineered Wound Healing Products

Tissue-engineered wound healing products can be classified as acellular or cellular. Acellular products contain, as the name implies, no cells and consist of a matrix that functions by binding to the host, allowing matrix-cell interactions. Due to its porous nature, the matrix allows host cells to infiltrate.Today, a matrix can contain virus vectors or plasmids, which can transcribe and translate the in-built DNA leading to secretion of specific growth factors. These growth factors carry out their specific functions by stimulating host cells to enhance wound healing. In specific situations it is even possible to manufacture matrices containing plasmids that can release hormones. The matrix contains ECM-proteins such as collagen, hyaluronic acid, and fibronectin, thus ensuring biocompatibility. Examples of these products are: E-Matrix™(Encelle Inc, Greenville, NC), OASIS® (Healthpoint Ltd, Fort Worth,Tex), Integra® (Integra LifeSciences, Plainsboro, NJ), Permacol® (Tissue Science Laboratories Inc,Andover, Md), Matriderm® (Dr. Suwelack Skin & Health Care AG, Germany), and EZDerm® (Brennen Medical, St. Paul, Minn).

Restoring function after hand burns plays a major role in the restitution of quality of life.The grafted areas are of utmost importance for good hand function. Haslik et al evaluated the collagen-elastin matrix Matriderm as a possible alternative for 2-stage substitutes after hand burns.This matrix was evaluated in several trials by van Zuijlen et al and showed the sufficient survival of autografts on top of the dermal substitutes in a 1-step procedure. Haslik et al concluded that Matriderm is a promising dermal substitute for the treatment of severe hand burns, with an overall take rate of 97%.

Cellular products on the other hand do contain living cells, often fibroblasts and keratinocytes embedded in a collagen or polyglactin scaffold forming an epidermal layer skin substitute. Autologous cells are used in these products to minimize the risk of rejection. Autologous keratinocytes are derived from progenitor cells from dermal sheets in the outer root surrounding hair follicles or from epithelial cells obtained via a biopsy of the recipient's skin. A permanent autologous epidermal skin graft is applied, which functions as a reliable barrier and promotes the formation of granulation tissue. Examples of tissue-engineered epidermal substitutes are: Epicel® (Genzyme Biosurgery, Cambridge, Mass), Laserskin® (Fidia Advanced Biopolymers, Abano Terme, Italy), Myskin ™ (Celltran Ltd, Sheffield, UK), and EpiDex ™ (Modex Therapeutics, Switzerland).

Laserskin autograft is an epidermal substitute. There are orderly arrays of laser-perforated microholes for the ingrowth and proliferation of keratinocyte. As keratinocytes are directly cultivated on Laserskin, the graft can easily be peeled off from the skin. Lam et al showed in an animal experiment that composite Laserskin grafts are good human skin substitutes in terms of durability, biocompatibility, high seeding efficacy for keratinocytes, high graft take rate, and low infection rate.

Aside from epidermal substitutes, tissue-engineered dermal substitutes have also been developed. The dermis is composed of loose connective tissue containing collagen and fibrils, anchoring the dermis to the epidermis. The upper layer of the dermis, the papillary layer, is a cell-rich layer containing fibroblasts and macrophages allowing dermis-epidermis interactions. These interactions trigger synthesis of ECM components and stimulation of differentiation and growth of keratinocytes. Inclusion of living fibroblasts leads to active release of cytokines. However, until now, only allogeneic tissue-engineered dermal wound products have been developed, and unfortunately always carry the risk of (chronic) graft rejection. Examples of these products include: Dermagraft® (Advanced BioHealing, La Jolla, Calif), Alloderm® (LifeCell Inc, Branchburg, NJ), TransCyte® (Advanced BioHealing, La Jolla, Calif), and ICX-SKN® (Intercytex, Cambridge, UK).

Dermagraft is a cryopreserved human fibroblastderived dermal substitute. Fibroblasts incorporated with Dermagraft secreted VEGF, PDGF, insulin-like growth factor I (IGF-I),colony stimulating factors (CSF), interleukins (IL), tumor necrosis factor (TNF), and TGF-β.

Alloderm is a dermal collagen matrix derived from banked human skin that is specially treated to remove most cellular components. Alloderm has been successvanfully used in the resurfacing of full-thickness burn wounds in combination with an ultra-thin autograft that replaces the epidermis.

TransCyte is a laboratory-grown extracellular matrix of allogeneic human dermal fibroblasts. Noordenbos et al reported on the safety and efficacy of TransCyte for treatment of partial-thickness burns. Wounds treated with TransCyte healed faster, with no infection and less hypertrophic scarring,compared to wounds treated with silver sulfadiazine. More recently, ICX-SKN, an autosynthesised human collagen-based extracellular matrix with human dermal fibroblasts vascularised during healing of acute surgical wounds and showed integration and persistence during the healing process.

Currently, (allogeneic) bilayered products are available— eg, Apligraf ® (Organogenesis, Canton, Mass) and OrCel® (OrCel International, New York, NY). Apligraf, a living bilayered skin substitute, is capable of regenerating tissue in response to an injury.This is apparent because of the good interaction between Apligraf and the wound, thus making it suitable for wide application.

OrCel is a bilayered cellular matrix in which normal human allogeneic skin cells (epidermal keratinocytes and dermal fibroblasts) are cultured in 2 separate layers into a type I bovine collagen sponge. Like Apligraf, OrCel is a bilayer dressing resembling normal skin. OrCel was developed as a tissue-engineered biological dressing. OrCel delivers ECM components and growth factors and creates an environment conducive to wound healing—it was never intended as an artificial skin for grafting.

EXTRACELLULAR MATRIX AND STRUCTURE

In biology, the **extracellular matrix** (**ECM**) is the extracellular part of multicellular structure (e.g., organisms, tissues, biofilms) that typically provides structural and biochemical support to the surrounding cells.Because multicellularity evolved independently in different multicellular lineages, the composition of ECM varies between multicellular structures; however, cell adhesion, cell-to-cell communication and differentiation are common functions of the ECM.

The animal extracellular matrix includes the interstitial matrix and the basement membrane. Interstitial matrix is present between various animal cells (i.e., in the intercellular spaces). Gels of polysaccharides and fibrous proteins fill the interstitial space and act as a compression buffer against the stress placed on the ECM.

Basement membranes are sheet-like depositions of ECM on which various epithelial cells rest.

The plant ECM includes cell wall components, like cellulose, in addition to more complex signaling molecules. Some single-celled organisms adopt multicelluar biofilms in which the cells are embedded in an ECM composed primarily of extracellular polymeric substances (EPS).

ROLE AND IMPORTANCE

Due to its diverse nature and composition, the ECM can serve many functions, such as providing support, segregating tissues from one another, and regulating intercellular communication. The extracellular matrix regulates a cell's dynamic behaviour.

In addition, it sequesters a wide range of cellular growth factors and acts as a local stores for them. Changes in physiological conditions can trigger protease activities that cause local release of such stores. This allows the rapid and local growth factor-mediated activation of cellular functions without *de novo* synthesis. Formation of the extracellular matrix is essential for processes like growth, wound healing, and fibrosis. An understanding of ECM structure and composition also helps in comprehending the complex dynamics of tumor invasion and metastasis in cancer biology as metastasis often involves the destruction of extracellular matrix by enzymes such as serine proteases, threonine proteases, and matrix metalloproteinases.

Molecular components

Components of the ECM are produced intracellularly by resident cells and secreted into the ECM via exocytosis. Once secreted, they then aggregate with the existing matrix. The ECM is composed of an interlocking mesh of fibrous proteins and glycosaminoglycans (GAGs).

Proteoglycans

GAGs are carbohydrate polymers and are usually attached to extracellular matrix proteins to form proteoglycans. Proteoglycans have a net negative charge that attracts positively charged sodium ions (Na^+), which attracts water molecules via osmosis, keeping the ECM and resident cells hydrated. Proteoglycans may also help to trap and store growth factors within the ECM.

Heparan sulfate

Heparan sulfate (HS) is a linear polysaccharide found in all animal tissues. It occurs as a proteoglycan (PG) in which two or three HS chains are attached in close proximity to cell surface or ECM proteins. It is in this form that HS binds to a variety of protein ligands and regulates a wide variety of biological activities, including developmental processes, angiogenesis, blood coagulation, and tumour metastasis. In the extracellular matrix, especially basement membranes, the multi-domain proteins perlecan, agrin, and collagen XVIII are the main proteins to which heparan sulfate is attached.

Chondroitin sulfate

Chondroitin sulfates contribute to the tensile strength of cartilage, tendons, ligaments, and walls of the aorta. They have also been known to affectneuroplasticity.

Keratan sulfate

Keratan sulfates have a variable sulfate content and, unlike many other GAGs, do not contain uronic acid. They are present in the cornea, cartilage, bones, and the horns of animals.

NON-PROTEOGLYCAN POLYSACCHARIDE

HYALURONIC ACID

Hyaluronic acid (or "hyaluronan") is a polysaccharide consisting of alternating residues of D-glucuronic acid and N-acetylglucosamine, and unlike other GAGs, is not found as a proteoglycan. Hyaluronic acid in the extracellular space confers upon tissues the ability to resist compression by providing a counteracting turgor (swelling) force by absorbing significant amounts of water. Hyaluronic acid is thus found in abundance in the ECM of load-bearing joints. It is also a chief component of the interstitial gel.

Hyaluronic acid is found on the inner surface of the cell membrane and is translocated out of the cell during biosynthesis. Hyaluronic acid acts as an environmental cue that regulates cell behaviour during embryonic development, healing processes, inflammation, and tumordevelopment. It interacts with a specific transmembrane receptor, CD44.

FIBERS

Collagen

Collagens are the most abundant protein in the ECM. In fact, collagen is the most abundant protein in the human body and accounts for 90% of bone matrix protein content. Collagens are present in the ECM as fibrillar proteins and give structural support to resident cells. Collagen is exocytosed inprecursor form (procollagen), which is then cleaved by procollagen proteases to allow extracellular assembly.

Disorders such as Ehlers Danlos Syndrome,osteogenesis imperfecta, and epidermolysis bullosa are linked with genetic defects in collagen-encoding genes. The collagen can be divided into several families according to the types of structure they form:

1. Fibrillar (Type I, II, III, V, XI)
2. Facit (Type IX, XII, XIV)
3. Short chain (Type VIII, X)
4. Basement membrane (Type IV)
5. Other (Type VI, VII, XIII)

Elastin

Elastins, in contrast to collagens, give elasticity to tissues, allowing them to stretch when needed and then return to their original state. This is useful in blood vessels, the lungs, in skin, and the ligamentum nuchae, and these tissues contain high amounts of elastins.Elastins are synthesised by fibroblasts andsmooth muscle cells. Elastins are highly insoluble, and tropoelastins are secreted inside a chaperone molecule, which releases the precursor molecule upon contact with a fibre of mature elastin. Tropoelastins are then deaminated to become incorporated into the elastin strand. Disorders such as cutis laxaand Williams syndrome are associated with deficient or absent elastin fibers in the ECM.

Fibronectin

Fibronectins are glycoproteins that connect cells with collagen fibers in the ECM, allowing cells to move through the ECM. Fibronectins bind collagen and cell-surface integrins, causing a reorganisation of the cell's cytoskeleton and facilitating cell movement.

Fibronectins are secreted by cells in an unfolded, inactive form. Binding to integrins unfolds fibronectin molecules, allowing them to form dimers so that they can function properly. Fibronectins also help at the site of tissue injury by binding to platelets during blood clotting and facilitating cell movement to the affected area during wound healing.

Laminin

Laminins are proteins found in the basal laminae of virtually all animals. Rather than forming collagen-like fibers, laminins form networks of web-like structures that resist tensile forces in the basal lamina. They also assist in cell adhesion. Laminins bind other ECM components such as collagens, nidogens, and entactins.

CELL ADHESION TO THE ECM

Many cells bind to components of the extracellular matrix. Cell adhesion can occur in two ways; by focal adhesions, connecting the ECM to actin filaments of the cell, and hemidesmosomes, connecting the ECM to intermediate filaments such as keratin.

This cell-to-ECM adhesion is regulated by specific cell-surface cellular adhesion molecules (CAM) known as integrins. Integrins are cell-surface proteins that bind cells to ECM structures, such as fibronectin and laminin, and also to integrin proteins on the surface of other cells.

Fibronectins bind to ECM macromolecules and facilitate their binding to transmembrane integrins. The attachment of fibronectin to the extracellular domain initiates intracellular signalling pathways as well as association with the cellular cytoskeleton via a set of adaptor molecules such as actin.

CELL TYPES INVOLVED IN ECM FORMATION

There are many cell types that contribute to the development of the various types of extracellular matrix found in plethora of tissue types. The local components of ECM determine the properties of the connective tissue.

Fibroblasts are the most common cell type in connective tissue ECM, in which they synthesise, maintain, and provide a structural framework; fibroblasts secrete the precursor components of the ECM, including the ground substance. Chondrocytes are found in cartilage and produce the cartilagenous matrix.Osteoblasts are responsible for bone formation.

Extracellular matrix in plants

Plant cells are tessellated to form tissues. The cell wall is the relatively rigid structure surrounding the plant cell. The cell wall provides lateral strength to resist osmotic turgor pressure, but it is flexible enough to allow cell growth when needed; it also serves as a medium for intercellular communication. The cell wall comprises multiple laminate layers of cellulose microfibrils embedded in a matrix of glycoproteins, including hemicellulose, pectin, and extensin. The components of the glycoprotein matrix help cell walls of adjacent plant cells to bind to each other. The selective permeability of the cell wall is chiefly governed by pectins in the glycoprotein matrix. Plasmodesmata (*singular*: plasmodesma) are pores that traverse the cell walls of adjacent plant cells. These channels are tightly regulated and selectively allow molecules of specific sizes to pass between cells.

Medical applications

Extracellular matrix cells have been found to cause regrowth and healing of tissue. In human fetuses, for example, the extracellular matrix works with stem cells to grow and regrow all parts of the human body, and fetuses can regrow anything that gets damaged in the womb. Scientists have long believed that the matrix stops functioning after full development. It has been used in the past to help horses heal torn ligaments, but it is being researched further as a device for tissue regeneration in humans.

In terms of injury repair and tissue engineering, the extracellular matrix serves two main purposes. First, it prevents the immune system from triggering from the injury and responding with inflammation and scar tissue.

Next, it facilitates the surrounding cells to repair the tissue instead of forming scar tissue.

For medical applications, the cells required are usually extracted from pig bladders, an easily accessible and relatively unused source. It is currently being used regularly to treat ulcers by closing the hole in the tissue that lines the stomach, but further research is currently being done by many universities as well as the U.S. Government for wounded soldier applications. As of early 2007, testing was being carried out on a military base in Texas. Scientists are using a powdered form on Iraq War veterans whose hands were damaged in the war.

Not all ECM devices come from the bladder. Extracellular matrix coming from pig small intestine submucosa are being used to repair "atrial septal defects" (ASD), "patent foramen ovale" (PFO) and inguinal hernia. After one year 95% of the collagen ECM in these patches is replaced by the normal soft tissue of the heart.

Extracellular matrix proteins are commonly used in cell culture systems to maintain stem and precursor cells in an undifferentiated state during cell culture and function to induce differentiation of epithelial, endothelial and smooth muscle cells in vitro.

Extracellular matrix proteins can also be used to support 3D cell culture in vitro for modelling tumor development.

A class of biomaterials derived from processing human or animal tissues to retain portions of the extracellular matrix are called ECM Biomaterial.

THE EXTRACELLULAR MATRIX OF NORMAL SKIN

The largest component of normal skin is the ECM, a gel-like matrix produced by the cells that it surrounds. The ECM is composed of a variety of polysaccharides, water and collagen proteins which give the skin remarkable properties. On a weight basis, the tensile (breaking) strength of normal skin approaches that of steel, yet skin also has substantial elasticity and compressibility. These properties are due to the combination of two main classes of ECM molecules, which are secreted by fibroblasts and epidermal cells. They are:

- Fibrous structural proteins, including collagens, elastin and laminin, which give the ECM strength and resilience
- Proteoglycans, such as dermatan sulfate and hyaluronan, typically consist of multiple glycosaminoglycan chains (formed from repeating disaccharide units) that branch from a linear protein core. Extracellular proteoglycans are large, highly hydrated molecules that help cushion cells in the ECM.

COLLAGEN

The largest class of fibrous ECM molecules is the collagen family, which includes at least 16 different types of collagen. Collagen in the dermal matrix is composed primarily of type I (80–85%) and type III (8–11%) collagens, both of which are fibrillar or rod-shaped collagens. The tensile strength of skin is due predominately to these fibrillar collagen molecules, which self-assemble into microfibrils in a head-to-tail and staggered side-to-side lateral arrangement. Collagen molecules become cross-linked to adjacent collagen molecules, creating additional strength and stability in collagen fibres.

Type IV collagen molecules associate with other specialised ECM molecules, including laminin and proteoglycans, to form a sheet-like structure known as the basement membrane (basal lamina), to which epidermal cells attach. This separates the epidermis from the dermis and surrounds blood vessels. Laminin is a large protein made up of three polypeptide chains that form a cross-shaped molecule. The four ends of the cross contain specialised regions that bind to type IV collagen molecules, heparan sulphate proteoglycan and specific proteins in the plasma membrane of cells called integrin receptors. This enables laminin to act as a multi-adhesive matrix protein by forming bridges between cells and the basement membrane.

Elastin

Many tissues, such as skin, lung and blood vessels, need to be both strong and elastic to function properly. A network of elastic fibres in the ECM of these tissues gives them the required resilience to recoil after stretching. The main component of elastic fibres is the elastin molecule, which creates cross-links to adjacent elastin molecules. These molecules form a core of elastic fibres and are covered by fibrillin, a large glycoprotein that binds to elastin and is essential for the integrity of elastic fibres.

Fibronectin

This is another important glycoprotein that is present in both plasma and tissue. Fibronectins have multiple functions, which allow them to interact with many extracellular substances, such as collagen, fibrin and heparin, and with specific membrane receptors on responsive cells.

Fibronectin has many different types of binding sites for collagen and fibrin molecules and heparan sulphate proteoglycan, allowing it to link together different types of ECM molecules. It also contains an important cell-binding domain made up of the three amino acids, Arg-Gly-Asp (RGD). Specific integrin receptors in the plasma membrane of cells recognise this RGD sequence of amino acids in fibronectin molecules. The binding of fibronectin molecules to integrin receptors on cells leads to the stimulation of signalling pathways that promote cell attachment, migration and differentiation. These characteristics enable fibronectin to play an important role in cell adhesion and to communicate signals between cells and components of the ECM.

Glycosaminoglycans

Glycosaminoglycans (GAGs) are composed of polysaccharide chains made up of repeating disaccharide units and are strongly hydrophilic. GAGs are highly negatively charged and therefore attract osmotically active Na+, causing large amounts of water to be drawn into their structure. This results in GAGs occupying a huge volume relative to their mass and forming gels at very low concentrations.

The hydrophilic nature of GAGs causes a swelling pressure, or turgor, which enables the ECM to withstand compression forces. The cartilage matrix lining the knee joint, for example, can support pressures of hundreds of atmospheres because of its high GAG content.

Hyaluronic acid (HA), a chief component of the ECM, is a large GAG that attracts water and is found in increased amounts in damaged or growing tissues. HA stimulates cytokine production by macrophages, thereby promoting angiogenesis.

Proteoglycans

Proteoglycans retain water and form a gel-like substance through which ions, hormones and nutrients can move freely. Given the great abundance and structural diversity of proteoglycan molecules, it would be surprising if their function was limited to providing lubrication around and between cells.

Perlecan is a proteoglycan present in the basement membrane, where it forms a gel of varying pore size and charge density. The epidermal cells of the skin derive their metabolic nutrients from the dermis via the diffusion of molecules through the basement membrane, so proteoglycans may play a key role in selectively filtering molecules that pass through the basement membrane beneath the epidermal cells.

Proteoglycans also appear to have an important role in regulating signalling between cells. For example, the heparin sulphate chains of proteoglycans bind to several different growth factors, including fibroblast growth factors (FGFs), helping them to bind to their specific cell-surface receptors.

Proteoglycans can also bind chemokine molecules on the surface of endothelial cells, prolonging the inflammatory response. Although growth factors typically bind to the GAG chains of proteoglycans, members of the transforming growth factor beta (TGF-β) family bind to the core protein of the decorin proteoglycan. TGF-β directly increases scar formation by increasing the expression of collagen, fibronectin and lysyl oxidase while reducing the expression of MMPs, which break down the ECM. TGF-β is also chemotactic for macrophages and neutrophils, which prolongs the inflammatory response and contributes to scarring. When TGF-β is bound by the core protein of decorin it is not able to interact with its normal receptor protein on target cells, which inhibits the activity of TGF-β. Proteoglycans such as syndecan protect elastase from the inhibition by alpha-1 proteinase inhibitor, suggesting that they modify the proteolytic environment of wounds.

EXTRACELLULAR MATRIX IN CHRONIC WOUNDS

All chronic wounds begin as acute wounds with a fibrin clot, but instead of progressing through the four phases of healing they become 'stuck' in a prolonged inflammatory phase. It has been proposed that this lengthened inflammatory phase causes increased levels of proteases such as MMPs, elastase, plasmin and thrombin, which destroy components of the ECM and damage the growth factors and their receptors that are essential for healing. Other factors that may contribute to the failure of some wounds to heal include elevated levels of oxygen free radicals, which chemically alter these essential components.

HISTOLOGICAL CHANGES IN CHRONIC WOUNDS

Non-healing chronic wounds

Herrick and colleagues investigated sequential changes in the location and levels of key proteins of the ECM, such as fibronectin, during the healing of chronic venous leg ulcers. They examined biopsies of the ulcer base, margin and surrounding skin before and after treatment with compression therapy and observed the following:

- Numerous small and medium-sized blood vessels were surrounded by prominent 'fibrin cuffs' containing concentric layered structures rich in fibrin, collagen and laminin
- The chronic ulcer base was covered with a layer of fibrinous exudate of variable thickness that contained numerous polymorphonuclear leukocytes
- Extravasated red blood cells, small perivascular deposits of haemosiderin and macrophages were present in the ulcer margin, base and surrounding skin
- Fibronectin staining markedly decreased at the base of the ulcer margins and was virtually absent from the central ulcer base and

fibrinous exudates, with the exception of the fibrin cuffs around the blood vessels. Immunostaining for fibronectin showed a diffuse pattern in the ECM of the surrounding skin, with a punctate staining pattern at the epidermal-dermal junction

- Collagen I and III were evident throughout the ECM of the surrounding skin. Within the ulcer base, bundles of collagen III were visible, lying parallel to the surface and differentiating between the fibrinous exudate and the surface of the granulation tissue. Collagen I staining was less evident in the ulcer base but was present in the basement membrane zone at the ulcer edge.

Healing chronic wounds

After two weeks of compression bandaging therapy, the following important histological changes were apparent in the venous leg ulcers:

- The epidermis of the surrounding skin was thicker than normal skin and highly keratinised
- The fibrinous exudate was thinner on the ulcer base and regenerating epidermis was visible
- Newly formed blood vessels and granulation tissue of variable maturity were present in the ulcer base and margins, and there was a marked increase in the number of polymorphonuclear leukocytes infiltrating the tissue. Macrophages had also increased in number
- Blood vessels cuffed with fibrin, laminin, fibronectin, collagen and tenascin were less abundant in both the ulcer edge and base
- Some red blood cell extravasation persisted, but there was a striking increase in the amount of haemosiderin deposited in the surrounding tissue
- The intensity and distribution of fibronectin staining throughout the ulcer base and edge was dramatically increased compared with the initial biopsy, and collagen III immunostaining was more uniform.

As healing progressed, epithelium covered the edges of the ulcers but was often detached from the underlying dermis. Reduced haemosiderin and focal red blood cell extravasation were observed in the surrounding skin and the ulcer margin. Angiogenesis was pronounced, with capillaries being surrounded by a thin layer of laminin consistent with the thickness of a normal basement membrane, and thicker fibrin cuffs were absent. A fine fibre matrix of fibronectin and collagen was present.

After complete epithelialisation, the pattern of fibronectin and collagen staining of the former ulcer base was indistinguishable from the adjacent surrounding tissue. However, the regenerated epidermis often became detached, suggesting weak attachment of the epidermal cells to the basement membrane. This histological study provides important insight into the molecular, cellular and structural abnormalities that are typical of chronic venous leg ulcers and the progression of changes that occur during healing. Other studies have compared the differences in wound fluid and ECM in acute wounds and chronic diabetic and venous ulcers. The predominant types of inflammatory cells in early wound healing are granulocytes, lymphocytes and macrophages. The main function of granulocytes in wounds is to eliminate contaminating bacteria.

The functions of lymphocytes in acute wound healing are less well established but they probably play a key role in activating macrophages, which are important in regulating the transition between wound inflammation and proliferation. Loots and colleagues found that chronic diabetic or venous ulcers contained significantly higher numbers of granulocytes than acute healing wounds on days five to 28 post-wounding. Levels of B cells, monocytes and macrophages were also significantly higher in chronic wounds than in acute wounds, especially 28 days after injury when the inflammatory response is generally decreasing in healing wounds. By contrast, the numbers of helper T cells were lower in chronic wounds than in acute wounds.

The location and levels of proteoglycans in normal skin, acute wounds and chronic wounds is just beginning to be investigated. An analysis of proteoglycans in venous leg ulcers and normal skin found some differences. In general, the proteoglycans in normal skin tended to be located at the plasma membrane surface of keratinocytes, while in venous ulcers the proteoglycans were mainly located in the keratinocyte cytoplasm. As some proteoglycan molecules bind growth factors and are necessary for their interaction with receptors, alterations in the location and levels of proteoglycans could influence growth factor systems in chronic wounds, justifying further investigations.

MOLECULAR ANALYSES OF ECM IN CHRONIC WOUNDS

Studies examining the molecular profiles of ECM in chronic wounds produced data that generally corroborated the histological changes observed in the later studies already discussed. One of the first reports to analyse the molecular status of ECM molecules in chronic wounds assessed the stability of fibronectin and vitronectin in fluids collected from chronic wounds. The results demonstrated that fibronectin in plasma (control) and healing surgical wounds was intact.

By contrast, fibronectin was totally degraded in the fluid collected from two venous stasis ulcers and was partially degraded in the exudate collected from two chronic diabetic foot ulcers. Importantly, the addition of intact plasma fibronectin to chronic wound fluids resulted in rapid degradation of the fibronectin. These results indicated that levels of intact soluble fibronectin were high in fluids from acute healing wounds but were dramatically reduced in fluids from some chronic wounds. Similar results were found for vitronectin, with significant degradation of vitronectin observed in chronic wound fluids. These results predated but further support the histochemical data of Herrick and colleagues, who found reduced immunostaining of fibronectin in chronic venous ulcers.

A subsequent study assessed the effect of chronic wound fluids on the attachment of fibroblasts to collagen-coated wells ; a high number of human dermal fibroblasts added to medium containing 10% serum attached to the collagen-coated wells. By contrast, reduced levels of cell attachment occurred when fluids from two chronic venous stasis leg wounds were added to the medium. Since fibronectin and vitronectin function as cell attachment and migration factors for many types of cells, degradation of these proteins by high protease activities probably contributes to the delays seen in epidermal resurfacing by keratinocytes, and poor granulation tissue formation by vascular endothelial cells and fibroblasts.

Data on elastin degradation in chronic skin wounds is lacking, but it is known that elastase is responsible for fibronectin degradation in burn wounds. It remains to be determined whether this also indicates that elastin degradation products can be found in chronic wounds. The most likely source of elastase in chronic wounds is from the polymorphonuclear leukocytes or macrophages that are present. Fibroblasts can also express elastase but its action appears to be directed at pre-elastic fibres, with limited activity towards mature dermal elastic fibres. Elastin degradation or its decreased expression leads to the loss of elastolytic properties in skin, for example with ageing or in scar tissue.

Evidence of collagen degradation in chronic wounds has largely been inferred from data measuring levels of MMPs in fluid obtained from chronic ulcers, rather than by analysing fluids directly for collagen breakdown products. Several studies have reported elevated levels of MMPs. Furthermore, the initial ratio of MMP-9/TIMP-1 in the exudate of chronic pressure ulcers inversely correlated with the healing of pressure ulcers. Analysis of biopsies of chronic pressure ulcers has also shown that, compared with normal skin tissue, MMP levels are highly elevated in pressure ulcers and that there is a strong association with inflammatory cells.

The immunostaining for MMP-1, MMP-2, MMP-3, and MMP-9 is very faint or absent in normal skin. In marked contrast, levels of all four MMPs are highly elevated in biopsies of chronic pressure ulcers. Furthermore, immunostaining for MMP-9 and MMP-2 is strongly associated with the inflammatory cells, which in these biopsies are present as a band of cells located in the superficial layers of the ulcer bed.

Other proteases may also play important roles in damaging ECM components or growth factors. The serine protease urokinase plasminogen activator (uPA) is known to activate a proteolytic cascade, resulting in the activation of MMPs. Analysis of fluids and biopsies from acute and chronic wounds found lower levels of active uPA and MMP-9 in acute wounds than in chronic venous leg ulcers. More importantly, there was a temporal switch from the active to the inactive form of uPA that coincided with a reduction in MMP-9 levels when chronic wounds began to heal. This result reinforces the importance of natural inhibitors of proteases, such as alpha-1 protease inhibitor and TIMPs, in the healing process. The lack of regulation of protease activity in chronic wounds results in degradation of the components of the ECM, leading to the eventual tissue breakdown, ulceration and skin stripping frequently seen in chronic wounds such as venous leg ulcers and pressure ulcers.

THERAPEUTIC INTERVENTIONS WITH THE POTENTIAL TO CORRECT DEGRADED ECM

These research finding have led to the development of several wound dressings with the aim of inactivating MMPs in chronic wounds. One of these, which consists of oxidised regenerated cellulose and collagen (Promogran®), has been shown to reduce levels of MMP activity and degrade exogenous PDGF in chronic wound fluid *in vitro*, presumably by acting as a competitive substrate for wound fluid proteases, which spares the PDGF at the expense of degrading the collagen in the dressing.

One clinical study suggests that this dressing may improve the healing of some chronic diabetic ulcers. A product that combines collagen with alginate (Fibracol®) has also been developed to absorb wound exudate (alginate component) and reduce MMP protease activity by providing a competitive substrate (collagen component).

A dressing containing a mixture of metal ions and citric acid (DerMax®) is reported to reduce oxygen free radicals and levels of MMP-2 *in vitro*, and small non-randomised studies are now under way to evaluate its use in diabetic foot ulcers and pressure ulcers.

Another new approach in development is to treat wounds topically with a unique protein that normally functions to enhance cell attachment during tooth development.

This protein, called amelogenin, is purified from porcine teeth and has been used with substantial success in about 750,000 patients undergoing periodontal procedures. Amelogenin is being evaluated for its ability to promote dermal wound healing based on its ability to provide a temporary ECM protein for cell attachment.

EXTRACELLULAR MATRIX IN ACUTE WOUNDS

Acute wounds normally heal in an orderly and efficient manner by progressing through four distinct but overlapping phases: haemostasis, inflammation, proliferation and remodelling. Throughout these phases, components of the ECM play an important role in regulating and integrating many key processes of healing.

HAEMOSTASIS AND THE PROVISIONAL WOUND MATRIX

The clot that forms at the site of an injury is necessary to stop bleeding. However, it plays other important roles in wound healing by depositing at the wound interface a host of plasma and cell-secreted constituents. The epidermal cells subsequently dissect their way under the clot and over the granulation tissue, which is comprised of a dense population of macrophages, fibroblasts and newly formed blood vessels embedded in a loose matrix of fibrin, fibronectin, collagen and other ECM proteins. Stimulation of the clotting cascade results in the proteolytic cleavage of fibrinogen by the enzyme thrombin, forming an insoluble fibrin clot that holds damaged tissues together and provides the provisional matrix. In addition, the clot contains fibronectin molecules that are present in plasma and bind to fibrin through fibrin-specific binding sites.

About three days after injury fibroblasts begin to express new integrin receptors. On the fourth day, fibroblasts migrate into the provisional wound matrix as part of angiogenesis and in response to chemotactic growth factors released by platelets from the periwound area and subcutaneous tissue. Using an in vitro model, fibronectin was found to be necessary for fibroblasts to express new integrin receptors and to migrate effectively over a collagen-fibrin matrix. The integrin receptors then generate new intracellular signals that stop the fibroblasts migrating. Growth factors and proteins contained within the provisional wound matrix help to stimulate fibroblasts to begin proliferating and synthesising new collagen and other ECM components.

In this way the provisional wound matrix functions as much more than an inert scaffold in which scar tissue is deposited. It acts as a reservoir to help trap growth factors and actively signals fibroblasts, epidermal cells and vascular endothelial cells, via their integrin receptors, to transform into activated wound cells that will repair the injury.

The inflammatory response

Neutrophils: These are the first inflammatory cells to respond to the soluble mediators released by platelets and the coagulation cascade. They begin their journey from the vasculature into the injured tissue by first adhering to the vascular endothelial cell walls via their surface integrin receptors.

They then release elastase and collagenase, which facilitate their migration through the basement membrane that surrounds the endothelial cells and into the ECM at the wound site. Their primary role is to mount the first line of defence against infection by phagocytosing and killing bacteria, and by breaking down foreign materials and devitalised tissue. Neutrophils also produce and release inflammatory mediators such as tumor necrosis factor alpha (TNF-α) and interleukin-1 (IL-1), which further recruit and activate fibroblasts and epithelial cells. Neutrophils produce and contain high levels of destructive proteases and oxygen free radicals, which they use to digest phagocytosed materials. On their inevitable death, neutrophils release these substances into the local wound area. This can cause extensive tissue damage and prolong the inflammatory phase. The persistent presence of high levels of bacteria in a wound may contribute to chronicity through continued recruitment of neutrophils and their release of proteases, cytokines and intracellular contents.

Monocytes/macrophages: Neutrophils are usually depleted in the wound after two to three days and are replaced by tissue macrophages. Macrophages begin as circulating monocytes that are attracted to the wound site, a process that begins about 24 hours after injury, by both soluble mediators and degraded components of the ECM, such as fragments of collagen and fibronectin. Monocytes bind to the ECM via integrin receptors and immediately differentiate into tissue macrophages. Serum factors and fibronectin mediate this differentiation.

Tissue macrophages have a dual role in the healing process. They are voracious phagocytes and patrol the wound area, ingesting bacteria, devitalised tissue and depleted neutrophils. Macrophages also produce collagenases and elastase to assist them in breaking down devitalised tissues. They are able to regulate proteolytic destruction of tissue in the wound by producing and secreting inhibitors for these enzymes.

Soluble mediators: Macrophages also mediate the transition from the inflammatory phase to the proliferative phase of healing, a role that is as important as their phagocytic role. They release a wide variety of growth factors and cytokines, including TNF-α, TGF-β, PDGFs, IL-1, interleukin 6 (IL-6), insulin-like growth factor-one (IGF-1) and FGF. Some of these soluble mediators recruit and activate fibroblasts, which will synthesise, deposit and organise the new tissue matrix, while others promote angiogenesis.

The proliferative phase

During the repair phase, the provisional wound matrix is remodelled and replaced with scar tissue, consisting of new collagen fibres, proteoglycans and elastin fibres, which partially restore the structure and function of the tissue. This is accomplished by the migration, proliferation and differentiation of epithelial cells, dermal fibroblasts and vascular endothelial cells from adjacent uninjured tissue and stem cells that originate in the bone marrow and circulate to the wound site.

Fibroblasts in the normal dermis are typically quiescent and sparsely distributed, in contrast to those in the provisional wound matrix and granulation tissue, which are numerous and active. Fibroblasts migrate into the wound in response to soluble cytokines and growth factors, which are initially released from platelets when they degranulate and later by macrophages in the wound. These include PDGF, TGF-β and basic FGF.

The direction of fibroblast movement is determined by the concentration gradient of chemotactic factors and by the alignment of the fibrils in the ECM and provisional matrix. Fibroblasts tend to migrate along these fibrils, as opposed to across them. They begin moving by binding first to matrix components such as collagen, fibronectin, vitronectin and fibrin via their cell-surface integrin receptors. While one end of the fibroblast remains bound to the matrix component, the cell extends a cytoplasmic projection to find another binding site. When the next site is found, the attachment to the original site is broken by proteases secreted by the fibroblast and the cell uses its cytoskeletal network of actin fibres to pull itself forward. These proteases are called matrix metalloproteinases (MMPs) and are essential for the migration of cells through the ECM.

The most important of these MMPs are collagenase (MMP-1), which cuts intact collagen at a single site; gelatinases (MMP-2 and MMP-9), which degrade partially denatured collagen (gelatin); and stromelysin (MMP-3), which degrades multiple protein substrates in the ECM. In addition, MMPs remove collagen and other ECM components that were denatured during the injury. This is important because the collagen molecules must interact very specifically with each other to properly form a collagen fibril. Partially degraded collagen molecules will not bind (interact) properly with new collagen molecules synthesised during scar formation, resulting in disorganised, weak ECM, so the degraded collagen molecules must be removed by controlled action of the MMPs. This is like remodelling a brick wall. When a hole is knocked in a brick wall to permit a new doorway or window, the end of the rows of bricks must be carefully prepared to enable the correct alignment of new bricks that are added to create the correct structure. MMPs 'chew back' the denatured matrix to reach intact functional matrix. However, this process must be carefully controlled by tissue inhibitors of metalloproteinases (TIMPs) to prevent the MMPs from degrading intact, functional matrix.

Fibrin peptides that have been degraded by proteolytic enzymes can increase vascular permeability and induce angiogenesis. Peptides of fibronectin, collagen and elastin can stimulate cell migration, influence cell proliferation and lead to arteriolar vasodilation. Thus, the controlled actions of proteases on ECM components play a key role in regulating angiogenesis and many other aspects of normal wound healing, including the migration of cells into the wound.

The process of angiogenesis is stimulated by local factors in the wound microenvironment, including low oxygen tension, low pH and high lactate levels. Several growth factors, including bFGF, TGF-β and vascular endothelial growth factor (VEGF), are also potent angiogenic signals for endothelial cells. It is now recognised that oxygen levels in the tissues directly regulate angiogenesis by interacting with oxygen-sensing proteins that regulate the transcription of angiogenic and anti-angiogenic genes. For example, the synthesis of VEGF by capillary endothelial cells is increased by hypoxia through the activation of the recently identified transcription factor hypoxia-inducible factor (HIF), which binds oxygen. When oxygen levels surrounding capillary endothelial cells drop, levels of HIF increase inside the cells. HIF-1 binds to specific DNA sequences and stimulates the transcription of specific genes, such as that encoding VEGF, that promote angiogenesis. The regulation of angiogenesis involves both stimulatory factors, such as VEGF, and anti-angiogenic factors, such as angiostatin, endostatin, thrombospondin and pigment epithelium-derived factor.

The cells that form the new capillaries were assumed to arise from pre-existing capillary endothelial cells, but it is clear from animal experiments that angiogenic factors cause endothelial cells adjacent to an ischaemic site to begin to migrate into the matrix and proliferate, forming new buds or sprouts. However, recent data suggest that some of the cells that form new capillaries actually come from bone marrow stem cells called haemangioblasts that circulate in the bloodstream. Furthermore, these bone marrow-derived cells have been found to contribute to both collagen type I and III, compared with the resident cell population that only transcribes type I collagen. Angiogenic factors are released as a result of either injury or ischaemia, causing haemangioblasts to migrate into the ECM and differentiate into capillary endothelial cells. The migration of these cells into the matrix requires the local secretion of proteolytic enzymes, especially MMPs. As the tip of a newly formed endothelial sprout comes into contact with another, a cleft develops; this subsequently becomes the lumen of the evolving vessel and a complete new vascular loop is formed. This process continues until the capillary system is sufficiently repaired and tissue oxygenation and metabolic needs are met. It is these new capillary loops embedded in a loosely deposited collagen matrix that give granulation tissue its characteristic uneven or granular appearance.

After the fibroblasts have migrated into the provisional wound matrix, they proliferate and begin to synthesise new collagen, elastin, proteoglycans and other components that comprise granulation tissue. PDGF and TGF-β are two of the important growth factors that regulate the expression of ECM genes and proteases in fibroblasts. Recent data indicate that a new growth factor, named connective tissue growth factor, mediates many of the effects of TGF-β on the synthesis of ECM.

Remodelling

This is the final phase of wound healing. The maturation of granulation tissue involves a reduction in the number of capillaries as smaller vessels are aggregated into larger ones, and a decrease in the amount of GAGs and proteoglycans. Cell density and metabolic activity in the granulation tissue decrease during maturation. Changes also occur in the type, amount and organisation of collagen, leading to an enhancement of the tensile strength of the tissues. Initially, type III collagen is synthesised at high levels, but this is replaced by type I collagen, the dominant fibrillar collagen in skin. The tensile strength of a newly epithelialised skin wound is only about 25% of normal tissue. Healed tissue is therefore never as strong as uninjured tissue. Tissue tensile strength is enhanced primarily by the reorganisation of collagen fibres, which are deposited randomly during granulation, and by increased covalent cross-linking of collagen molecules by the enzyme lysyl oxidase, which is secreted into the ECM by fibroblasts. Over several months or more, changes in collagen organisation in the repaired tissue will slowly increase the tensile strength to a maximum of about 80% of normal tissue.

Remodelling of the ECM proteins occurs through the actions of several different classes of proteolytic enzymes produced by cells in the wound bed at different times during the healing process. Two of the most important families are MMPs and serine proteases. Specific MMP proteases that are necessary for wound healing are the collagenases, which degrade intact fibrillar collagen molecules; the gelatinases, which degrade damaged fibrillar collagen molecules; and the stromelysins, which degrade proteoglycans. An important serine protease is neutrophil elastase, which can degrade almost all types of protein molecules. Under normal conditions, the destructive actions of proteolytic enzymes are carefully regulated by specific enzyme inhibitors, which are produced by cells in the wound bed. The TIMPs specifically inhibit MMPs.

INFLAMMATION PROCESS

The 'inflammatory process' includes a tissue-based startle reaction to trauma; go/no-go decisions based on integration of molecular clues for tissue penetration by microbes; the beckoning, instruction and dispatch of cells; the killing of microbes and host cells they infect; liquefaction of surrounding tissue to prevent microbial metastasis; and the healing of tissues damaged by trauma or by the host's response.

If at any step an order to proceed is issued but progress to the next step is blocked, the inflammatory process may detour into a holding pattern, such as infiltration of a tissue with aggregates of lymphocytes and leukocytes (granulomas) that are sometimes embedded in proliferating synovial fibroblasts (pannus), or distortion of a tissue with collagen bundles (fibrosis). Persistent inflammation can oxidise DNA badly enough to promote neoplastic transformation.

What Celsus defined around AD40 as 'rubor, calor, dolor, tumor' (redness, heat, pain and swelling) is today an intellectually engaging problem in signal transduction and systems biology, as well as a multibillion dollar market for the pharmaceutical industry. When primary pathogenetic events are unknown, control of inflammation is sometimes the next best option. The number of diseases considered 'inflammatory' in origin may decline as infectious causes continue to be discovered for some of them, such as *Helicobacter pylori*-dependent chronic gastritis with ulcer formation. However, in this and several other important infectious diseases, the inflammatory response may cause more damage than the microbe. Although the search continues for possible infectious causes of multiple sclerosis, rheumatoid arthritis and atherosclerosis, inflammation *per se* remains one of the main therapeutic targets in diverse disorders with a staggering collective impact.

Inflammation is usually life preserving, as reflected by the increased risk of grave infections in people with genetic deficiencies in principal components of the inflammatory process. For example, inability to mobilise leukocytes to sites of inflammation in type I or II leukocyte adhesion deficiency, if untreated, often leads to death from infection. Inability to produce the complement components properdin and factors D, C5, C6, C7, C8 or C9 predisposes to meningococcal infection. Thus, the medical focus on inhibiting inflammation is accompanied by an effort of potentially comparable importance to learn how to induce inflammation more effectively, in at least two important settings. First, causing and prolonging inflammation are among the essential functions of adjuvants, and a better understanding of the role of inflammation in adjuvanticity may enable prophylactic immunisation against a wider range of infectious diseases.

Second, generation of inflammation is one of the main goals of tumour immunology, both for therapeutic immunisation and for nonspecific immunostimulation, such as by instilling Bacille Calmette-Guérin into the urinary bladder to prevent recurrence of tumours.

The accompanying articles in this issue integrate cross-sections of inflammation biology by peering inside blood vessels, joints, brain, viscera and epithelia.

The papers form a backdrop against which to evaluate diverse new anti-inflammatory treatments. These include neutralisers of tumour-necrosis factor (TNF); blockers of leukotriene receptors; inhibitors of cyclooxygenase (COX)-2, leukotriene synthetase and 3-hydroxy-3-methylglutaryl coenzyme A reductase; and agonism at protease-activated receptor 1 by activated protein C. Many more anti-inflammatory compounds are in the pipeline.

In this article I offer a perspective on inflammation as a system of information flow in response to injury and infection. If tissue is injured, the basic challenge facing the host is to detect whether there is accompanying infection. If infection is the initial event, the challenge is to detect whether tissue is injured.

When injury and infection coincide, the goal is to react as quickly as possible to terminate the spread of infection, even at the cost of further tissue damage. The need to detect two states at once before risking self-inflicted damage dictates a dependence on binary or higher-order signals. The need to accelerate at a potentially high cost brings with it the need to decelerate as soon as the goal has been met. A full stop requires repairing the tissue whose damage triggered inflammation or that inflammation damaged.

Such a complex system can be characterised by its checkpoints. I first consider checkpoints evident early and late after an inflammatory response is activated, and then present evidence that another set of checkpoints operates constitutively in the basal state to prevent the inappropriate initiation of inflammation.

GO SIGNALS IN EARLY CHECKPOINTS

Evolution did not anticipate surgery with aseptic technique. Thus, the body reacts to trauma as if the emergency is infection, until proven otherwise.

The take-home message is apparent with the following experiment. Expose one forearm with the inner surface facing up. Spread the three middle fingers on your other hand and slap them down hard on your forearm. Within about 15 seconds the skin of your forearm will display a red bas-relief of the offending digits.

Over the next hour the image will fade. In contrast, if the epidermis had been broken and bacteria had entered, redness and swelling would persist, testifying to an escalating series of events that is synchronised according to bacterial replication time and metastatic potential. The episode would probably culminate in the confinement and killing of the penetrant bacteria and the destruction and repair of a small amount of tissue. Then again, if the inflammatory response were feeble and antibiotics unavailable, the outcome might be death from sepsis.

The flow of information following mild trauma with infection. Tissue damage unleashes up to three types of go signals. First, in response to pain, neurons release bioactive peptides. Second, broken cells release constitutively expressed intracellular proteins that trigger cytokine production when found in the extracellular space.

Examples include heat-shock proteins, the transcription factor HMGB1 (for high mobility group 1) and mitochondrial peptides bearing the *N*-formyl group characteristic of prokaryotic proteins. Third, microbes and their shed or secreted products are sensed through binding of their conserved molecular constituents to soluble receptors such as complement, mannose-binding protein and lipopolysaccharide-binding protein, and to cell-surface receptors such as Toll family members, peptidoglycan recognition proteins and scavenger receptors.

Much attention in inflammation research has focused on the recruitment of leukocytes from the blood. However, a rapid response requires sentinel cells pre-stationed in the tissues. Mast cells and macrophages fulfil this function. The importance of mast cells as first responders, recently emphasised in experimental rheumatoid arthritis, is symbolised by their placement atop. Responding to the signals, perivascular mast cells release histamine, eicosanoids, pre-formed TNF, newly synthesised cytokines, tryptases, other proteases, and chemokines. Histamine, eicosanoids and tryptases cause vasodilatation (responsible for the heat and redness) and extravasation of fluid (the cause of swelling).

Mast-cell tryptases cleave protease-activated receptors whose neo-termini then engage G-protein-coupled receptors on mast cells, sensory nerve endings, endothelium and neutrophils. This further activates mast cells and neurons, makes endothelium sticky for leukocytes and leaky to fluid, and prompts leukocytes to release platelet-activating factor (PAF). PAF reinforces the pro-adhesive conversion of endothelium, which results in leukocyte emigration from the vasculature. For simplicity, interactions among endothelial cells, leukocytes and extravascular signals are omitted. The impacts of the coagulation and kinin cascades on interactions of endothelium and leukocytes and the reciprocal influence of inflammation on the interactions of endothelium and coagulation factors.

Neutrophils are partially activated (primed) by the TNF and leukotrienes produced by mast cells and by other neutrophils, leading to release of small amounts of elastase. This cleaves the anti-adhesive coat of CD43 (leukosialin) from neutrophils, allowing their integrins to engage extracellular matrix proteins.

The binary signal of integrin engagement plus stimulation by TNF, chemokines or C5a triggers degranulation and a massive respiratory burst, resulting in release of proteinases (such as the serprocidins elastase, cathepsin G and protease 3), other hydrolases, antibiotic proteins (such as bacterial permeability increasing factor, four -defensins, the three serprocidins and their proteolytically inactive homologue, azurocidin) and oxidants (such as hydrogen peroxide, hypohalites and chloramines). The oxidants activate matrix metalloproteinases (MMPs) and inactivate protease inhibitor.

The foregoing actions promote tissue breakdown. Metalloproteinases cleave TNF from tissue macrophages as well as from monocytes that are chemotactically attracted from the bloodstream into the tissue by azurocidin. Macrophage- and monocyte-derived TNF and chemokines attract and activate more neutrophils.

TNF and chemokines combine with mast cell-derived prostaglandin E2 (PGE2) and neutrophil-derived defensins to recruit lymphocytes, while leukotrienes help attract antigen-presenting dendritic cells. Lymphocytes, in conjunction with microbial products, activate macrophages to secrete proteases, eicosanoids, cytokines and reactive oxygen and nitrogen intermediates (ROIs and RNIs, respectively).

In summary, the inflammatory system is geared for lag-free acceleration, but requires ongoing verification of emergency to avoid defaulting to the resting state. Each newly recruited cell generally commits to release pro-inflammatory signals only after integrating inputs of both host and microbial origin.

It is a canon of immunology that for cellular activation, B cells generally need antigen-receptor engagement plus signals from T cells; T cells need antigen-receptor engagement plus signals from antigen-presenting cells (APCs); and APCs, including macrophages, need cytokines plus microbial products, or cytokines plus CD40 ligation, or microbial products plus products of necrotic host cells. The stresses that a requirement for binary or higher-order go signals begins with the activation of mast cells and neutrophils, and that sustained activation of mast cells and neutrophils usually precedes and conditions the activation of APCs, T cells and B cells as the inflammatory response evolves into the immune response. That a combination of tissue injury plus infection sustains inflammation helps clarify what provokes an immune response.

Massive trauma, post-ischaemic or toxic necrosis, and haemorrhage and resuscitation can each trigger an inflammatory response that appears to be independent of infection.

This may reflect the ability of some host cell products that are altered (for example, fragmented matrix proteins or oxidised lipoproteins), abnormally released (for example, heat-shock proteins) or released in abnormally large amounts to interact with receptors (for example, Toll-like receptor 4) that otherwise detect microbial signals. Alternatively, cryptic microbial signals may be involved, because such stresses may be associated with the translocation of bacteria or diffusion of their products across the intestinal wall.

STOP SIGNALS IN EARLY CHECKPOINTS

Superimposed on the feed-forward cycles are sets of brakes. Brakes involving lipid autacoids illustrate one mechanism: to progressively raise the threshold for continuing the inflammatory reaction. Neutrophil-derived arachidonate serves as substrate for neutrophil 5-lipoxygenase to generate the inflammatory leukotriene B4. However, as neutrophils infiltrate tissues, they also pass arachidonate to tissue cells expressing 15-lipoxygenase, which produces lipoxins. Lipoxins are a class of oxidised eicosanoids that bind cellular receptors and block neutrophil influx. Neutrophils also pass to other cells a 5-lipoxygenase intermediate, leukotriene A4; 15-lipoxygenase converts this to a lipoxin as well.

In this manner, cell–cell interactions favour a transition in the profile of arachidonate products from pro-inflammatory leukotrienes to anti-inflammatory lipoxins. At the same time, COX2 is induced in macrophages by microbial products and cytokines. COX2 converts arachidonate to PGE2, which contributes to fluid leak from blood vessels. However, as PGE2 levels rise, PGE2 feeds back to inhibit COX2 as well as 5-lipoxygenase, while transcriptionally inducing 15-lipoxygenase in neutrophils. These delayed effects shift arachidonate metabolism towards lipoxin formation in neutrophils themselves. In this way, over several hours, PGE2, at first a go signal, becomes a stop signal. The anti-inflammatory drug aspirin recapitulates this phenomenon by acetylating COX2; the acetylated enzyme switches from making PGE2 to making lipoxins.

Studies with gene-disrupted mice highlight additional stop signals. Mice lacking the ectonucleotidase CD39 over-react to chemical irritation of the skin. Mice deprived of purinergic A2a receptors succumb to normally sublethal doses of microbial and chemical toxins. These observations suggest that CD39 breaks down extracellular ATP and ADP secreted by activated cells or leaking from broken cells, generating adenosine. Adenosine then acts to suppress inflammatory responses by neighbouring cells.

the cell-surface immunoglobulin

In another set of examples, mice lacking the cell-surface immunoglobulin superfamily molecule CD200 suffer more macrophage influx and worse experimental autoimmune encephalomyelitis and collagen-induced arthritis than do wild-type mice. Similarly, mast cells lacking the integrin-binding receptor gp49B1 degranulate excessively in response to immunoglobulin E–antigen complexes.

These studies hint at a wide array of protein–protein interactions among cells, and between cells and their matrix, that temper inflammation in its early phase.

A fourth type of stop signal is issued by the autonomic nervous system. As reviewed by Tracey in this issue , cholinergic discharge blocks the release of TNF from macrophages in the viscera.

THE INFLAMMATORY PROCESS AND RESPONSE

Inflammation, the adaptive immune response to tissue injury or infection, plays a central role in metabolism in a variety of organisms.

At its most basic level, an acute inflammatory response is triggered by 1) tissue injury (trauma, exposure to heat or chemicals); or 2) infection by viruses, bacteria, parasites, or fungi.

The classic manifestation of acute inflammation is characterised by four cardinal signs: Redness and heat result from the increased blood flow to the site of injury. Swelling results from the accumulation of fluid at the injury site, a consequence of the increased blood flow. Finally, swelling can compress nerve endings near the injury, causing the characteristic pain associated with inflammation. Pain is also important to make the organism aware of the tissue damage. Additionally, inflammation in a joint usually results in a fifth sign (impairment of function), which has the effect of limiting movement and forcing rest of the injured joint to aid in healing.

A well-controlled acute inflammatory response has several protective roles:

- It prevents the spread of infectious agents and damage to nearby tissues;
- helps to remove damaged tissue and pathogens, and;
- assists the body's repair processes

However, a third type of stimuli, **cellular stress and malfunction,** triggers *chronic inflammation*, which, rather than benefiting health, contributes to disease and age-related deterioration via numerous mechanisms.

CELLULAR STRESS & CHRONIC, LOW-LEVEL INFLAMMATION

Mitochondria – cellular organelles responsible for generating biochemical energy in the form of*adenosine triphosphate* (ATP) – are a fundamentally necessary component of life in higher organisms. In fact, in the case of sophisticated multicellular life forms, organismal viability depends upon optimal mitochondrial function. Paradoxically, mitochondrial processes can also bring about a tissue-destroying inflammatory mediator known as *the inflammasome;* this phenomenon is provoked by damaged and dysfunctional mitochondria.

Mitochondrial dysfunction arises consequent of exposure to exogenous (e.g. environmental toxins, tobacco smoke) and endogenous (e.g. reactive oxygen species) stressors, and as a result of the aging process itself. For example, a byproduct of mitochondrial energy generation is the creation of *free radical molecules.* Free radicals can damage cellular structures and initiate a cascade of proinflammatory genetic signals that ultimately results in cell death (apoptosis), or worse, uncontrolled cell growth - the hallmark of cancer.

Aging is associated with declining mitochondrial efficiency and increased production of free radical molecules. Recent research identifies this age-associated aberration of mitochondrial function as a principle actuator of chronic inflammation. Specifically, mitochondrial dysfunction brings about inflammation as follows:

1. Accumulation of free radicals induces mitochondrial membrane permeability;
2. Molecular components normally contained within the mitochondria leak into the *cytoplasm*(intracellular fluid in which cellular organelles are suspended);
3. Cytoplasmic *pattern recognition receptors (PRR's),* which detect and initiate an immune response against intracellular pathogens, recognise the leaked mitochondrial molecules as potential threats;
4. Upon detection of the potential threat, PRR's form a complex called

the inflammasome that activates the inflammatory cytokine *interleukin-1β*, which then recruits components of the immune system to destroy the "infected" cell.

These four steps represent a simplified scheme of mitochondrial dysfunction leading to cellular destruction; however, intracellular free radicals are not the only inducers of inflammatory cell death. Circulating sugars, primarily *glucose* and *fructose*, are culprits as well. When these "blood sugars" come in contact with proteins and lipids a damaging reaction occurs forming compounds calledadvanced glycation end products (AGEs). AGEs bind to the cell-surface receptor called *receptor for advanced glycation end products*, or RAGE. Upon activation, RAGE triggers the movement of the inflammatory mediator *nuclear factor kappa-B* (NF-kB) to the nucleus, where it activates numerous inflammatory genes. Advanced glycation end products are primarily formed *in vivo,* and glycation is exacerbated by elevated blood sugar levels. However, dietary AGEs also contribute to inflammation; they are abundant in foods cooked at high temperatures, especially red meat.

Additional biochemical inducers of a chronic inflammatory response include:

- Uric acid (urate) crystals, which can be deposited in joints during gouty arthritis; elevated levels are a risk factor for kidney disease, hypertension, and metabolic syndrome;
- Oxidised lipoproteins (such as LDL), a significant contributor to atherosclerotic plaques ; and
- Homocysteine, a non-protein-forming amino acid that is a marker and risk factor for cardiovascular disease, and may increase bone fracture risk.

Together, these proinflammatory instigators promote a perpetual low-level chronic inflammatory state called para-inflammation.

Although it progresses silently, para-inflammation presents a major threat to the health and longevity of all aging humans. Chronic, low-level inflammation is associated with common diseases including cancer, type II diabetes, osteoporosis, cardiovascular diseases, and others. Thus, by targeting the myriad physiological variables that can inaugurate an inflammatory response, one can effectively temper chronic inflammation and reduce their risk for inflammatory diseases.

MARKERS AND MEDIATORS OF INFLAMMATION

The most prominent markers of inflammation used in research and diagnosis. Some can be detected by blood tests:

Tumor necrosis factor alpha (TNF-α) is an intercellular signaling protein called a cytokine, which can be released by multiple types of immune cells in response to cellular damage, stress, or infection. Originally identified as an anti-tumor compound produced by macrophages (immune cells) , TNF-α is required for proper immune surveillance and function. Acting alone or with other inflammatory mediators, TNF-α slows the growth of many pathogens. It activates the bactericidal effects of neutrophils, and is required for the replication of several other immune cell types. Excessive TNF-α, however, can lead to a chronic inflammatory state, can increase thrombosis (blood clotting) and decrease cardiac contractility, and may be implicated in tumor initiation and promotion.

Nuclear factor kappa-B (NF-kB) is important in the initiation of the inflammatory response. When cells are exposed to damage signals (such as TNF-α or oxidative stress), they activate NF-kB, which turns on the expression of over 400 genes involved in the inflammatory response. These include other inflammatory cytokines, and pro-inflammatory enzymes including *cyclooxygenase-2* (COX-2) and *lipoxygenase*. COX-2 is the enzyme responsible for synthesising pro-inflammatory prostaglandins, and is the target of non-steroidal anti-inflammatory drugs (ibuprofen, aspirin) and COX-2 inhibitors (Celebrex®).

Interleukins are cytokines that have many functions in the promotion and resolution of inflammation. Pro-inflammatory interleukins that have been the subject of most research include IL-1β, IL-6, and IL-8. IL-1β helps immune cells to move out of blood vessels and into damaged or dysfunctional tissues. IL-6 has both pro-inflammatory and anti-inflammatory roles, and coordinates the production of compounds required during the progression and resolution of acute inflammation. IL-8 is expressed by both immune and non-immune cells, and helps to attract neutrophils (immune cells that can destroy pathogens) to sites of injury.

C-reactive protein (CRP) is an acute-phase protein, one of several proteins rapidly produced by the liver during an inflammatory response. Its primary goal in acute inflammation is to coat damaged cells to make them easier to recognise by other immune cells. CRP elevation above basal levels is not diagnostic on its own, as it can raise in several cancers, rheumatologic, gastrointestinal, and cardiovascular conditions, and infections. Elevation of CRP (as determined by a high-sensitivity CRP assay or hs-CRP) has a strong association with elevated risk of cardiovascular disease and stroke (Emerging Risk Factors Collaboration et al. 2010).

Eicosanoids. The cytokine factors mentioned above (interleukins, TNF-α) are "long-distance messages". They are produced by cells at the site of inflammation and released into the blood, carrying information about the inflammatory response throughout the body. In contrast, eicosanoids are "local" messages; they are produced by cells that are proximal to the site of inflammation, and are meant to travel short distances (locally within the same organ, to neighboring cells, or sometimes only to different parts of the same cell) in order to elicit immune defenses.

There are several families of eicosanoids (including prostaglandins, prostacyclins, leukotrienes, and thromboxanes) that are created by most cell types in all major organ systems. Aside from their roles in inflammation (and anti-inflammation), prostaglandins have a variety of functions in cell growth, kidney function, digestion, and the constriction and dilation of blood vessels.

Thromboxanes are important mediators of the blood clotting process. Pro-inflammatory leukotrienes are important for recruiting and activating white blood cells during inflammation, and are best studied for their role in airway constriction and anaphylaxis.

Cells produce eicosanoids using unsaturated fatty acids that are part of their cell membranes. The fatty acid starting materials for eicosanoid synthesis are the essential fatty acids linoleic acid (omega-6) and its derivative arachidonic acid (AA); and alpha-linolenic acid (an omega-3) and its derivatives eicosapentaenoic acid (EPA) and *docosahexaenoic* acid (DHA). While generalisations about roles of these fatty acids in eicosanoid synthesis should be approached cautiously, the most potent inflammatory eicosanoids are produced from omega-6 fatty acids (linoleic and arachidonic acids). Diets high in omega-3 fatty acids are associated with lower biomarkers of inflammation and cardiovascular disease risk; proposed mechanisms include the production of less inflammatory or anti-inflammatory eicosanoids and through the cyclooxygenase and lipoxygenase enzymes.

Cyclooxygenases and Lipoxygenases. The eicosanoids require several enzymatic steps to be synthesised from unsaturated fatty acids; the cyclooxygenase (COX) and lipoxygenase (LOX) enzymes catalyse the first steps in these reactions. Cyclooxygenases initiate the conversion of omega-3 and omega-6 derivatives into one of the many prostaglandins or thromboxanes. The interest in COX enzyme metabolism comes from the fact that its inhibition leads to decreased prostaglandin synthesis, and therefore a reduction in inflammation, fever, and pain.

The analgesic and anti-inflammatory activity of aspirin and the non-steroidal anti- inflammatory drugs (NSAIDS, like ibuprofen and naproxen) is due to their inhibition of COX enzymes. There are two COX enzymes with well-defined roles in humans (COX-1 and COX-2). COX-2 has the most relevance to the inflammatory process: it is normally inactive, but is turned on during inflammation and stimulates this process of inflammation by creating pro-inflammatory prostaglandins and thromboxanes.

Lipoxygenases convert fatty acids into proinflammatory *leukotrienes*, important local mediators of inflammation. Several potent inflammatory leukotrienes are produced by 5-LOX in mammals. Lipoxygenase enzymes, and the pro-inflammatory factors they produce, have a fundamental role in the inflammatory process by aiding in the recruitment of white blood cells to the site of inflammation.

They also stimulate local cells to produce cytokines, which amplifies the inflammatory response. Thus, LOX enzymes may be involved in a wide variety of inflammatory conditions, and represent an additional target for anti-inflammatory therapy

While COX and LOX enzymes are most often associated with pro-inflammatory processes, it is important to remember that both enzymes also produce factors that inhibit or resolve inflammation and promote tissue repair (including the prostacyclins and lipoxins), The proper transition from the pro- to anti-inflammatory activities of the COX and LOX enzymes is an important for the progression of a healthy inflammatory response.

RISK FACTORS FOR CHRONIC INFLAMMATION

There are several risk factors which increase the likelihood of establishing and maintaining a low-level inflammatory response. These include:

Age. In contrast to younger individuals (whose levels of inflammatory cytokines typically increase only in response to infection or injury), older adults can have consistently elevated levels of several inflammatory molecules, especially IL-6 and TNF-α. These elevations are observed even in healthy older individuals. While the reasoning for this age-associated increase in inflammatory markers is not thoroughly understood, it may reflect cumulative mitochondrial dysfunction and oxidative damage, or may be the result of other risk factors associated with age.

Obesity. Fat tissue is an endocrine organ, storing and secreting multiple hormones and cytokines into circulation and affecting metabolism throughout the body. For example, fat cells produce and secrete both TNF-α and IL-6, and visceral (abdominal) fat can produce these inflammatory molecules at levels sufficient to induce a strong inflammatory response. Visceral fat cells can produce three times the amount of IL-6 as fats cells elsewhere , and in overweight individuals, may be producing up to 35% of the total IL-6 in the body.

Fat tissue can also be infiltrated by macrophages, which secrete pro-inflammatory cytokines. This accumulation of macrophages appears to be proportional to BMI, and appear to be a major cause of low-grade, systemic inflammation and insulin resistance in obese individuals.

Diet. A diet high in saturated fat is associated with higher pro-inflammatory markers, particularly in diabetic or overweight individuals. This effect was absent in healthy individuals. Diets high in synthetic trans- fats (such as those produced by hydrogenation) have been associated with increases in inflammatory markers (IL-6, TNF-α, IL-8, CRP) in some studies , but had no effect in others.

The increases in markers of inflammation due to synthetic trans- fats may be more pronounced in individuals that are also overweight.

General dietary over-consumption is a major contributor to inflammation and other detrimental age-related processes in the modern world. Therefore, eating a calorie-restricted diet is an effective means of relieving physiologic stressors.

Indeed, several studies show that calorie restriction provides powerful protection against inflammation. For more information about the metabolic benefits of eating fewer calories, readers should refer to the caloric restriction protocol. Low sex hormones. Amongst their many roles in biology, sex hormones also modulate the immune/inflammatory response. The cells that mediate inflammation (such as neutrophils and macrophages) have receptors for estrogens and androgens that enable them to selectively respond to sex hormone levels in many tissues.

The macrophages that reside in skeletal tissue and are responsible for breaking down and recycling old bone.

Estrogens turn down osteoclast activity. Following menopause, lowered estrogen levels cause these bone depleting cells to maintain their activity, breaking down bone faster than it is rebuilt. This is one of the factors in the progression of osteoporosis.

Experiments in cell culture have demonstrated that testosterone and estrogen can repress the production and secretion of several pro-inflammatory markers, including IL-1β, IL-6, TNF-α, and the activity of NF-kb.

These observations have been corroborated by observational studies that have linked lower testosterone levels in elderly men to increases in inflammatory markers (IL-6 and IL-6 receptor). Several studies have shown an increase in inflammatory IL-1β, IL-6, and TNF-α following surgical or natural menopause. Conversely, the preservation of sex hormone levels is associated with reductions in the risk of several inflammatory diseases, including atherosclerosis, asthma in women, and rheumatoid arthritis in men.

Hormone replacement therapy (HRT) may partially exert its protective effects through an attenuation of the inflammatory response. Reductions in the risks of coronary heart disease and inflammatory bowel disease in some individuals, as well as levels of some circulating inflammatory cytokines (including IL-1B, IL-8, and TNF-α) has been observed in some studies of women on hormone replacement therapy.

Smoking. Cigarette smoke contains several inducers of inflammation, particularly reactive oxygen species. Chronic smoking increases production of several pro-inflammatory cytokines (TNF-α, IL-1β, IL-6, IL-8), while simultaneously reducing production of anti-inflammatory molecules. Smoking also increases the risk of periodontal disease, an independent risk factor for increasing systemic inflammation.

Sleep Disorders. Production of inflammatory cytokines (TNF-α and IL-1β) appears to follow a circadian rhythm and may be involved in the regulation of sleep in animals and humans. Disruption of normal sleep can lead to daytime elevations of these pro-inflammatory molecules. Plasma levels of TNF-α and/or IL-6 were elevated in patients with excessive daytime sleepiness, including those with sleep apnea and narcolepsy. These elevations in cytokines were independent of body mass index or age, although persons with higher visceral body fat were more likely to have sleep disorders.

OTHER INCITING FACTORS

Periodontal disease can produce a systemic inflammatory response that may affect several other systems, such as the heart and kidneys. It is by this mechanism that periodontal disease is thought to be a risk factor for cardiovascular diseases

Stress (both physical and emotional) can lead to inflammatory cytokine release (IL-6); stress is also associated with decreased sleep and increased body mass (stimulated by release of the stress hormone cortisol), both of which are independent causes of inflammation.

The maintenance of a proper inflammatory response may also involve the central nervous system. The recently identified vagal immune reflex senses inflammatory molecules through a network of nerves (branches of the vagus nerve), and sends this information to the brain. If the brain determines that the inflammatory response is too great, it sends signals to the site(s) of inflammation to attenuate the response. Preliminary data suggests that depressed nerve activity may be associated with exaggerated inflammatory responses seen in sepsis. Smoking, itself a risk factor for inflammation, also decreases activity of the vagus nerve.

4

Diseases Associated with Chronic Inflammation

Cardiovascular diseases (CVD). Inflammation is an integral part of atherosclerosis (recall that oxidised low-density lipoprotein cholesterol stimulates the inflammatory response). Circulating inflammatory cytokines are predictive of peripheral arterial disease, heart failure, atrial fibrillation, stroke, and coronary heart disease.

Cancer. Several studies have established links between chronic low-level inflammation and many types of cancer, including lymphoma, prostate, ovarian, pancreatic, colorectal and lung. There are several mechanisms by which inflammation may contribute to carcinogenesis, including alterations in gene expression, DNA mutation, epigenetic alterations, promotion of tumor vascularisation, and the expression of pro-inflammatory cytokines that have roles in cancer cell proliferation.

Diabetes. The infiltration of macrophages into fat tissue and their subsequent release of pro-inflammatory cytokines into circulation occur at a greater rate in type II diabetics than in non-diabetics. Pro-inflammatory cytokines clearly decrease insulin sensitivity.

Age-related macular degeneration (AMD). An evaluation of 11 population-based studies encompassing over 41,000 patients demonstrated a clear association between elevated serum CRP levels (> 3 mg/L) and the incidence of late onset AMD. The risk of AMD in these high-CRP patients was increased over 2-fold compared with patients with CRP levels < 1 mg /L.

Chronic kidney disease (CKD). The chronic, low-grade inflammation in CKD can lead to the retention of several pro-inflammatory molecules in the blood (including cytokines, AGEs, and homocysteine). The reduced excretion of pro-inflammatory factors by the diseased kidney can accelerate the progression of chronic inflammatory disturbances elsewhere in the body, such as the cardiovascular system.

Osteoporosis. Inflammatory cytokines (TNF-α, IL-1β, IL-6) are involved in normal bone metabolism. Osteoclasts, the cells that break down (resorb) bone tissue, are a type of macrophage and can be stimulated by pro-inflammatory factors. Systemic elevations in pro-inflammatory cytokines push bone metabolism towards resorption, and have been observed to induce bone loss in persons with periodontal disease, pancreatitis, inflammatory bowel disease, and rheumatoid arthritis (Cao 2011). An increase in the levels of inflammatory cytokines is also a mechanism by which menopause stimulates bone loss.

Depression. There is a small, but significant association between elevated IL-6 and CRP in depressed patients, which has been observed in many population studies. It is unclear whether inflammation leads to stress or vice versa, and there is data supporting both hypotheses.

Cognitive decline. Several observational studies have linked chronic low-level inflammation in older adults to cognitive decline and dementia, including vascular dementia and Alzheimer's disease. One study found that people with the highest CRP and IL-6 levels (> 2.4 pg/mL) had a ~30-40% increased risk of cognitive decline compared to those with the lowest levels (< 1.4 pg/mL).. Inflammatory markers can be elevated before the onset of cognitive dysfunction, indicating their potential relevance as a prognostic tool in high-risk individuals. Others. Elevations in circulating inflammatory cytokines are associated with several other conditions, both inflammatory (rheumatoid arthritis, IBD/Crohn's disease, pancreatitis) and non-inflammatory (anemia, fibromyalgia, frailty, sacropenia/cachexia/muscle wasting). Again, whether inflammation incites these conditions or results from them is unclear, and requires further investigation.

CONVENTIONAL MEDICINE TYPICALLY OVERLOOKS CHRONIC INFLAMMATION

Chronic inflammation or para-inflammation is generally not treated on its own by mainstream physicians. Interventions in conventional medicine are usually only undertaken when the inflammation occurs in association with another medical condition (such as arthritis).

Currently, conventional preventive medical approaches to inflammation are limited to the use of CRP to predict cardiovascular disease in high-risk subjects, and the prophylactic use of drugs like aspirin to inhibit the inflammatory cascade linked to thrombosis (uncontrolled blood clotting). Indeed, the potentially asymptomatic nature of low grade inflammation is such that elevations of pro-inflammatory cytokines may progress undetected for some time, only being discovered after they have had time to cause enough cellular damage to produce disease symptoms. As future studies solidify the association between inflammatory mediators and different diseases, early detection of cytokine aberrations and anti-inflammatory therapy to reduce disease risk may gain more mainstream acceptance.

DRUG STRATEGIES TO COMBAT CHRONIC INFLAMMATION

Pentoxifylline. Pentoxifylline is a drug used to treat conditions involving poor circulation to the brain, limbs, and other areas perfused by small blood vessels. The drug effectively modulates properties of both blood vessels *and* red blood cells thanks to its action as a non-selective*phosphodiesterase* inhibitor. Phosphodiesterase inhibition is a clinically important mechanism in many additional aspects of human physiology as well, so pentoxifylline has been studied in a wide range of applications ranging from diabetic complications and non-alcoholic liver disease, to endometriosis and cardiac surgery.

The potent anti-inflammatory properties of pentoxifylline were a secondary discovery, and still are not fully understood. Studies have revealed, though, that pentoxifylline modulates TNF-α signaling, which probably contributes to the considerable suppression of inflammation it has evoked in several human trials. In a recent trial, 400 mg of pentoxifylline taken twice daily significantly suppressed hs-CRP, fibrinogen, and TNF-α levels in patients with chronic kidney disease; subjects renal function improved with treatment as well. In patients with HIV-related vascular dysfunction, pentoxifylline lessened leukocyte adhesion – a process that contributes to cardiovascular disease by allowing inflammatory cells to infiltrate the endothelial lining of blood vessels. Given by IV-infusion, pentoxifylline lowered TNF-α levels and pain intensity following surgical removal of kidney stones.

Pentoxifylline dosage varies depending on individual circumstances and clinical application, however, 400 mg taken twice daily has consistently tempered inflammation in diverse human trials. For example, administered at this dose for one month to 30 diabetic individuals with high blood pressure, not only did pentoxifylline quell inflammation (20% reduction in CRP levels and an 11% improvement in erythrocyte sedimentation rate [measure of inflammatory tendency of a blood sample]), but it also bolstered plasma antioxidant status, as evidenced by a 20% reduction in malondialdehyde levels (measure of oxidative stress) and a nearly 5% increase in glutathione levels, a powerful antioxidant.

Metformin. The regulation of energy metabolism and inflammation are closely associated; this is evidenced by the co-incidence of metabolic disorders (obesity, diabetes) and low-grade inflammation. Metformin may reduce the activity of inflammatory cytokines by increasing the production of IL-

1βreceptor *antagonist (IL1Rn)*, a protein factor which interferes with pro-inflammatory signaling of IL-1β. It may also promote favorable CRP levels, although not to the same extent as weight loss. A randomised controlled trial of hypertensive and dyslipedemic patients taking 1700 mg/day of metformin for 12 weeks demonstrated a 26.7% reduction in IL-6 and 8.3% reduction in TNF-α from baseline levels, a degree of reduction similar to that of the potent statin drug rosuvastatin (Crestor®). The anti-inflammatory effects of metformin appear to be rapid; reductions in circulating TNF-α, IL-1β, CRP, and fibrinogen were observed after only 30 days in a larger study of 128 type II diabetic patients with dyslipidemia.

Aspirin. Aspirin has been used as an anti-inflammatory therapy long before the molecular mechanics of inflammation had been discovered; it is now well characterised as an inhibitor of cyclooxygenase enzymes. The modification of COX molecules by aspirin has important implications for cardiovascular health. Blood platelets use cyclooxygenase to produce thromboxane A2, a pro-inflammatory molecule that is an important signal during the initial stages of the clotting process. The inhibitory effect of aspirin on COX enzymes in platelets can partially explain its protective effects against the complications of several disorders, including hypertension, heart attack, and stroke. Aspirin's inhibition of cycloxygenase also helps explain its potential effect on cancer risk reduction as observed in several studies, as COX-2 also appears to have roles in increasing the proliferation of mutated cells, tumor formation, tumor invasion, and metastasis, and may contribute to drug resistance in some cancers. Aspirin has also been shown to reduce the activity of NF-kb in vitro , and lower levels of multiple inflammatory markers (TNF-α,CRP, IL-6) in patients with cardiovascular disease.

Unlike many other non-steroidal anti-inflammatory drugs (NSAIDS), the effects of aspirin on COX enzymes are permanent for the life of the COX enzyme. Interestingly, it appears that rather than rendering the enzyme inactive, aspirin modifies the function of COX. Aspirin stops the enzyme from producing pro-inflammatory prostaglandins, and enables it to begin producing anti-inflammatory molecules called resolvins.

Low-Dose Statin Drugs. Statins are thought to reduce inflammation by a mechanism distinct from their effects on cholesterol metabolism; they interfere with the function of cytokine receptors on the surface of white blood cells. Therefore, pro-inflammatory signals in the blood are unable to provoke a response from white blood cells, and they are prevented from further stimulating inflammation. Results of the JUPITER trial presented the strongest evidence for statins as anti-inflammatory therapy; in this study of over 17,000 healthy middle-aged men and women with elevated levels of the inflammatory marker CRP but normal levels of blood lipids, 20mg/day of rosuvastatin

(Crestor®) reduced CRP levels by over half, in addition to reducing heart attack and stroke incidence. Smaller studies have looked at the effect of statins on other inflammatory markers as well. A randomised controlled trial of hypertensive and dyslipedemic patients taking a lower dose (10 mg/day) of rosuvastatin for 12 weeks demonstrated a ~22% reduction in IL-6 and 13% reduction in TNF-α from baseline levels. A second uncontrolled study of simvastatin demonstrated more modest reductions in IL-6, but no changes in TNF-α from the statin treatment. To generate a substantial anti-inflammatory effect using statin drugs alone requires a high dose that is more likely to induce side effects than lower dose statin therapy.

DIETARY APPROACHES TO REDUCE CHRONIC INFLAMMATION

Inflammation itself is not a disease, but is featured, to varying degrees, in adverse health conditions. Information on strategies and research regarding the reduction of inflammation characteristic to specific health conditions are featured in their respective Life Extension Protocols: Allergies; Age-related Macular Degeneration; Cancer Adjuvant therapy; Cardiovascular Disease; Gout; Inflammatory Bowel Disease; Osteoarthritis and Rheumatoid Arthritis; Osteoporosis. What follows is a summary of dietary and supplemental approaches to addressing general chronic inflammation and para-inflammation. As many types of general inflammation often occur without additional symptoms, most of the strategies listed below are based on their ability to reduce circulating inflammatory cytokines, the hallmark of the para-inflammatory state.

Macronutrients and Energy Balance. Macronutrient content (particularly the types and levels of carbohydrates and fats) can have a significant effect on the progression of inflammation (as measured by increases in pro-inflammatory markers).

Diets with relatively high glycemic index (GI) and glycemic load (GL) have been associated with elevated risk of coronary heart disease, stroke, and type 2 diabetes mellitus, particularly among overweight individuals, and have been associated with modest increases in proinflammatory markers in multiple studies. In a study of over 18,000 healthy women e”45 years old without diagnosed diabetes, high GI and GL diets resulted in a small but significant increase in hs-CRP (+12% for high GI) over low GI diets. In the Danish Hoorne study , for every 10 unit increase in dietary glycemic index, circulating CRP was increased by 29%. Some dietary fats (particularly saturated and synthetic trans- fats) increase inflammation occurrence, while omega-3 polyunsaturated fats appear to be anti-inflammatory.

Since fat tissue (especially abdominal fat) expresses inflammatory cytokines, obesity can be a major cause of low-grade, systemic inflammation. Thus, it is important that total energy intake be proportional to energy expenditure, to avoid the deposition of abdominal fat. Obesity-induced increases in inflammatory cytokines appear to be reversible with fat loss. In a dramatic example, weight loss (by adjustable gastric banding) in a group of 20 severely obese individuals reduced IL-6 by 22% and CRP by almost half.

An inflammatory index, developed by a group from the Arnold School of Public Health at the University of South Carolina, scored 42 common dietary constituents based on their ability to raise serum CRP. Constituents (such as saturated fat, tea polyphenols, or vitamin D) were given either a positive (anti-inflammatory) or negative (pro-inflammatory) score, the magnitude of which was weighted based on the volume of inflammation research on the isolated ingredient.

Human clinical data was weighted more than animal data, and clinical trials more than observational studies. The scores were then verified by comparing them to nutrient intakes and CRP levels from a group of 494 volunteers over the course of 1 year. Amongst the most anti-inflammatory nutrients (based on the model and study data) are magnesium, beta-carotene, turmeric (curcumin), genistein, and tea; the most pro-inflammatory included carbohydrates, total- and saturated fat, and cholesterol. The index may provide a useful metric for accessing the overall inflammatory potential of an individual diet.

Exercise. Energy expenditure through exercise lowers multiple cytokines and pro-inflammatory molecules independently of weight loss. While muscle contraction initially results in a pro-inflammatory state, it paradoxically lowers systemic inflammation. This effect has been observed in dozens of human trials of exercise training in both healthy and unhealthy individuals across many age groups.

Fibre. In an analysis of 7 studies on the relationship between weight loss and hs-CRP, increased fibre consumption correlated with significantly greater reductions in hs-CRP concentrations. In these studies, daily fibre intakes ranging from 3.3 to 7.8 g/MJ (equivalent to about 27 to 64 g/day for a standard 2000 kcal diet) reduced CRP from 25%-54% in a dose-dependent fashion. These results should be interpreted carefully, as only two of the seven studies were specifically designed to examine the effects of fibre independently. The Women's Health Initiative failed to detect an effect of fibre consumption on hs-CRP, but found that greater intake of dietary soluble and insoluble fibre (over 24 g/day) was associated with lower levels of IL-6 and TNF-α.

MICRONUTRIENTS

Magnesium. In two large observation studies (the Women Health Initiative and Harvard Nurses Study), greater magnesium (Mg) intake was associated with lower hs-CRP, IL-6, and TNF-α receptor, a measure of TNF-α activity (Galland 2010, Chacko et al. 2010). Data from the Multi-Ethnic Study of Atherosclerosis failed to find significant differences in IL-6 or CRP levels between individuals with the highest and lowest magnesium intakes, but did find a significant association between greater dietary magnesium and the lower levels of the inflammation-associated proteins homocysteine and fibrinogen. Magnesium was rated as the most anti-inflammatory dietary factor in the Dietary Inflammatory index, which rated 42 common dietary constituents on their ability to reduce CRP levels based on human and animal experimental and observation data.

Vitamin D. Vitamin D appears to exert anti-inflammatory activity by the suppression of pro-inflammatory prostaglandins, and inhibition of the inflammatory mediator NF-êβ. Although intervention studies of its anti-inflammatory activity in humans are lacking, several observational studies suggest vitamin D deficiency may promote inflammation. Vitamin D deficiencies are more common amongst patients with inflammatory diseases (including rheumatoid arthritis, inflammatory bowel disease, systemic lupus erythematosus, and diabetes) than in healthy individuals.

They also occur more frequently in populations that are prone to low-level inflammation, such as obese individuals and the elderly. Vitamin D levels can drop following surgery (a condition associated with acute inflammation), with a concomitant rise in CRP.

Low vitamin D status was associated with elevated CRP in a study of 548 heart failure patients , and with increases in IL-6 and NF-êβ in a group of 46 middle-aged men with endothelial dysfunction.

Vitamin E. Vitamin E functions as an antioxidant in the body. Specifically, vitamin E is incorporated into low-density lipoprotein (LDL) particles and protects them against oxidative damage; it seems to guard against atherosclerosis via other mechanisms as well. The gamma-tocopherol form of vitamin E appears to complement the anti-inflammatory action of alpha-tocopherol.

Gamma-tocopherol has been shown to inhibit COX-2 and attenuate IL-1β signaling. In a small clinical trial on subjects with metabolic syndrome, the combination of gamma-tocopherol and alpha-tocopherol effectively suppressed C-reactive protein and TNA-α levels compared to placebo. In this study, the combination of both tocopherols performed better than either alone, prompting the investigators to remark *"the combination of [alpha-tocopherol] and [gamma-tocopherol] supplementation appears to be superior to either supplementation alone on biomarkers of oxidative stress and inflammation and needs to be tested in prospective clinical trials..."*

Zinc and Selenium. Zinc- and Selenium-containing antioxidant proteins (such as superoxide dismutase and glutathione peroxidase) reduce reactive oxygen species (free radicals), which indirectly inhibits NF-êβ activity and prevents the production of several inflammatory enzymes and cytokines. Zinc can also inhibit NF-êβ in a more direct manner. Zinc supplementation is associated with decreases in inflammation in populations that are prone to zinc deficiency, such as children and the elderly. Low level inflammation and circulating pro-inflammatory factors (CRP, TNF-α, IL-6, and IL-8) were reduced in elderly subjects by moderate zinc supplementation in several studies.

Like zinc, selenium deficiencies are common in chronic inflammatory states associated with disease (such as sepsis) , where selenium supplementation has been associated with reductions in inflammation and better patient outcomes.

OTHER DIETARY FACTORS

Resveratrol and Pterostilbene. The exact mechanism by which resveratrol exerts anti-inflammatory activity has not been established, although it inhibits a variety of pro-inflammatory compounds (cyclooxygenase, TNF-α, IL-1β, IL-6, NF-êβ) in animal models and human cell culture. The related compound pterostilbene has demonstrated similar inhibition of inflammatory markers in cell culture. Modulation of the inflammatory immune response likely contributes to resveratrol's protective role in animal models of heart disease, cancer, acute pancreatitis and inflammatory bowel disease. Resveratrol may be protective against general, low-level para-inflammation as well: when taken with a single high-fat, high-carbohydrate meal (930 kcal), resveratrol (100 mg) prevented the sharp post-meal increases in markers of oxidation and inflammation in a small crossover study of 10 healthy volunteers. For example, synthesis of IL-1β increased by 91% over 5 hours following the test meal; with resveratrol, this increase was significantly less (29%).

Curcumin. Extensive in vitro and animal studies have examine the effects of curcumin on experimentally-induced inflammatory diseases (atherosclerosis, arthritis, diabetes, liver disease, gastrointestinal disorders, and cancers) and disease markers (lipoxygenase, cyclooxygenase, TNF-α, IL-1β, NF-êβ, and others). Fewer human studies have examined curcumin effects on patient-oriented outcomes in inflammatory diseases, but most of the small randomised controlled trials of curcumin have consistently shown patient improvements in several inflammatory diseases, including psoriasis, irritable bowel syndrome, rheumatoid arthritis, and inflammatory eye disease.

Tea polyphenols. The anti-inflammatory effects of green and black tea polyphenols have been substantiated by dozens of in vitro and animal studies The polyphenols EGCG and theaflavin exert their anti-inflammatory effects through the inhibition of the NF-êβ signaling pathway, which decreases expression of several inflammatory proteins (lipoxygenase, cyclooxygenase, TNF-α, IL-1β, IL-6, and IL-8) in cell culture experiments. EGCG also inhibits the production and release of histamine, a key mediator of allergic and inflammatory response, in vitro.

In observational studies of tea consumption, >2 cups of tea/day (black or green) was associated with a nearly 20% reduction in CRP compared to non-tea drinkers, and significantly lower levels of two other inflammatory markers (serum amyloid A and haptogen, which are elevated in coronary heart disease). In clinical interventions, black tea appears to be more successful in reducing inflammatory markers than green. A 25% reduction in CRP was also observed in a small trial of healthy, non-smoking men consuming a black tea extract (equivalent to 4 cups of tea/day) for 6 weeks. A similar average reduction was observed in a larger study of healthy, individuals at high risk for coronary heart disease, but revealed a more dramatic 40-50% reduction in CRP amongst individuals with the highest starting CRP values (>3 mg/L). Carotenoids.

In the Women's Health and Aging Study, participants with the highest blood levels of α-carotene and total carotenoids were significantly more likely to have the lower IL-6 levels than participants with low carotenoid levels at the onset of the study. Participants with the lowest blood levels of α- and β-carotene, lutein/zeaxanthin, or total carotenoids were more likely to experience increases in IL-6 over a period of 2 years.

DHEA. Low levels of sex hormones are associated with systemic increases in inflammatory markers ; DHEA (dehydroepiandrosterone) an adrenal steroid hormone, the precursor to the sex steroids testosterone and estrogen. DHEA is abundant in youth, but steady declines with advancing age and may be partially responsible for age-related decreases sex steroids. In cell culture and animal models, DHEA can suppress inflammatory cytokine activity, in some cases more effectively than either testosterone or estrogen. Chronic inflammation may itself reduce DHEA levels. DHEA supplementation in elderly volunteers (50 mg/day for 2 years) significantly decreased TNF-α and IL-6 levels, as well as lowered visceral fat mass and improved glucose tolerance (both associated with inflammation) in a small study.

Fish Oil, is the best source of the omega-3 fatty acids eicosapentaenoic acid — EPA, and docosahexaenoic acid – DHA that can only be synthesised to a limited extent in humans, which is why fish oil supplementation is so critical. Omega-3 fatty acids have been well studied for their prevention of cardiovascular disease and mortality in tens of thousands of patients; the anti-inflammatory effects of omega-3's contribute to this activity. They have also proven successful at improving patient outcomes in scores of studies of other inflammatory diseases, particularly asthma, inflammatory bowel disease, and rheumatoid arthritis.

The association between greater fish oil/omega-3 consumption and reduced systemic inflammation is substantiated by data from several large observational trials. In 855 healthy participants from the Health Professionals Follow-Up Study, intake of omega-3 fatty acids was associated with lower plasma levels of markers of TNF-α activity; interestingly, high intake of both omega-3 and omega-6 fatty acids (which are usually assumed to be pro-inflammatory) was associated with the lowest level of inflammation. The Nurses. Health Study I cohort of 727 women revealed lower concentrations of inflammatory markers (including CRP and IL-6) amongst those in the top 20% of omega-3 consumption, when compared to those who consumed least amount. In the ATTICA study of over 3000 Greek men and women without any evidence of cardiovascular disease, participants who consumed over 300 g of fish per week had, on average, 33% lower CRP, 33% lower IL-6, and 21% lower TNF-α than participants who did not consume fish. In a sample of 5,677 men and women without cardiovascular disease from the Multi-Ethnic Study of Atherosclerosis (MESA) cohort, long-chain omega-3 intake (from fish or supplements) was associated with reduced plasma concentrations of multiple inflammatory markers (including CRP, IL-6, and TNF-α receptor, a measure of TNF-α activity) N-acetyl cysteine (NAC). Activation of the NF-êB pathway plays a central role in the activation of inflammatory cytokine genes; N-acetyl cysteine inhibits NF-êβ in cell culture, lowering expression of cytokines such as IL-6 and IL-8. Data establishing the effects of NAC on lowering chronic inflammation in humans is limited, but shows promise. NAC supplementation for 8 weeks demonstrated modest, but statistically significant decreases in circulating IL-6 levels in patients with chronic kidney disease. The effects were more pronounced in persons with significant inflammation at the start of the study (as measured by hs-CRP). NAC also reduced markers of systemic inflammation in a small study of patients with burn injuries.

Boswellia. Boswellia serrata (frankincense) is a traditional anti-arthritic in Ayurvedic medicine; its anti-inflammatory properties have been attributed to the specific inhibition of 5-LOX and reduction in the production of pro-inflammatory leukotrienes by boswellic acids, a constituent of the Boswellia gum resin. In cell culture, both crude and highly purified Boswellia extracts inhibited the production of pro-inflammatory TNF-α and IL-1β. One of the boswellic acids, *Acetyl-11-keto-beta-boswellic acid (AKBA)*, was an inhibitor of NF-Kb activity in mice , while a topical mixture of the four most abundant boswellic acids decreased inflammation in a rodent inflammation model. A recent systematic review of human trials of Boswellia for inflammatory conditions revealed that the small number of randomised controlled trials on the extract have produced encouraging results for its use for asthma and osteoarthritis , warranting larger studies to confirm the extract as an effective therapy. Standardised Boswellia extracts (30% AKBA) have been effective in mitigating pain in osteoarthritis patients ; when combined with non-volatile Boswellia oil, the standardised extract (called AprèsFlex™, or Aflapin®) demonstrated improved activity at a lower concentration. The use of Boswellia extracts for inflammatory bowel diseases has been investigated in multiple clinical trials, although results have been mixed.

Sesame Lignans. The observation that sesame oil could decrease the production of arachidonic acid in fungi and rat liver cells led to the identification of the sesame lignans (sesamin, sesamolin, sesaminol) as specific inhibitors of Ä5 desaturase (delta-5-desaturase), one of the enzymes used in the synthesis of arachidonic acid. By inhibiting Ä5 desaturase, sesame lignans may reduce the synthesis of pro-inflammatory prostaglandin, leukotrienes, and thromboxanes, each of which require arachidonic acid as a starting material. In animal models, diets high in sesame seed oil reduced production of the pro-inflammatory prostaglandins PGE-1 and -2, as well as thromboxane B2. In humans, 5 weeks of sesamin supplementation (39 mg/day) reduced the production of the pro-inflammatory vasoconstrictor 20-hydroxyeicosatetraenoic acid (20-HETE; a product of the enzyme 5-LOX) by 30%. This potential anti-inflammatory property of sesame lignans may partially explain its observed hypotensive (blood pressure-lowering) activity.

Bromelain. The anti-inflammatory activity of the proteolytic enzyme preparation bromelain has been attributed to its ability to reduce COX-2 activity, decrease prostaglandin and thromboxane synthesis, lower circulating fibrinogen levels, and reduce cellular adhesion of pro-inflammatory white blood cells to the sites of inflammation. Human trials of bromelain for inflammatory conditions have yielded promising results. In a blinded study from Germany, researchers divided 90 patients with painful osteoarthritis of the hip into two groups: one half receiving an oral enzyme preparation containing bromelain for six weeks, while the other half received the anti-inflammatory drug diclofenac (sold under the brand name Voltaren® and generic names). They found that the bromelain preparation was as effective as diclofenac in standard scales of pain, stiffness and physical function, and better tolerated than the drug comparator. The researchers concluded, "[the bromelain preparation] may well be recommended for the treatment of patients with osteoarthritis of the hip with signs of inflammation as indicated by a high pain level".

Another study comparing a standardised commercial enzyme preparation containing bromelain with diclofenac reached the same conclusion. The study reported that the supplement containing bromelain (90 mg, three times daily) to be as effective as diclofenac (50 mg, twice daily) in improving the symptoms of osteoarthritis of the knee. Patients reported comparable reductions in joint tenderness, pain and swelling, and improvement in range of motion at the end of the study. The investigators found bromelain to be as good as diclofenac on a standard pain assessment scale and to be better than the drug in reducing pain at rest (by 41% for bromelain versus 23% for the drug), improving restricted function (by 10% for bromelain versus 0% for the drug), being rated by more patients in improving symptoms (24% for bromelain versus 19% for the drug), and being evaluated by more physicians as having good efficacy (51% for bromelain versus 37% for the drug). In summary, the investigators determined bromelain to be an effective and safe alternative to NSAIDs such as diclofenac for painful osteoarthritis.

In further research from the United Kingdom, a three-month study looked at the dose-dependent effects of bromelain, either 200 mg or 400 mg a day in volunteers with mild acute knee pain. Pain evaluation was based on patient symptom scores, which were reduced by 41% in the 200 mg bromelain group and by 59% in those receiving 400 mg of bromelain, indicating a dose-response relationship. This was also observed for scores of stiffness and physical function, which decreased significantly in the higher-dose bromelain group compared with those receiving 200 mg. The researchers also noted that overall psychological well-being was significantly improved in both bromelain groups, leading to their conclusion that this natural therapy may be effective in improving general well-being as well as symptoms in otherwise healthy adults suffering from mild knee pain.

In animal models and cell culture experiments, bromelain has consistently demonstrated a variety of anti-inflammatory properties.

MITOCHONDRIAL SUPPORT

Reactive oxygen species generated during mitochondrial respiration contribute to inflammation. Aging individuals are especially susceptible to mitochondria-related oxidative stress since mitochondria become increasingly dysfunctional with age. Taking steps to support mitochondrial integrity and efficiency can help alleviate some of the systemic oxidative and inflammatory burden caused by poorly functioning mitochondria. Two nutrients, coenzyme Q10 (CoQ10) and pyrrloquinoline quinone (PQQ) are powerful mitochondrial protectants , and studies support an anti-inflammatory role for these compounds.

Pyrroloquinoline quinone is a cofactor for enzymes critically important for cellular energy homeostasis and redox balance. Several studies have shown that PQQ exerts a protective effective during circumstantial mitochondrial stress and increased oxidative load. In one study, rats given a diet supplemented with PQQ displayed greater energy expenditure and, remarkably, increased mitochondrial density in liver tissue. PQQ supplemented rats also had lower triglycerides and their hearts were more protected against lack of oxygen than rats that had not been given PQQ. During periods of limited oxygen supply to cardiac tissue, a dramatic spike in oxidative stress and subsequent inflammation damages cells; the findings from this animal model indicate that PQQ can stave off this inflammatory cell destruction by preserving mitochondrial efficiency in adverse conditions.

Coenzyme Q10 is an indispensable intermediary in mitochondrial ATP production. Studies have shown that CoQ10 levels are low during inflammatory conditions. In one investigation, patients with septic shock were found to have CoQ10 levels substantially lower than healthy individuals, and, among patients, lower CoQ10 levels correlated with higher levels of an inflammatory mediator called VCAM. In an animal model in which rats were given drinking water with added fructose, an experiment that leads to obesity, diabetes, and other inflammatory complications, CoQ10 supplementation attenuated the inflammatory response by decreasing hepatic expression of CRP and other inflammatory mediators.

Laboratory experiments indicate that CoQ10 modulates the expression of several hundred genes, many involved in inflammatory signaling. Of particular significance, one experiment showed that CoQ10, at physiologically relevant concentrations, was able to blunt induced TNF-α by more than 25% via modulation of the NF-kB signaling pathway.

GUARDING AGAINST INFLAMMATORY GLYCATION REACTIONS

The role of elevated blood sugar and glycation end products in initiating an inflammatory storm. Fortunately, in addition to reducing caloric intake to suppress both fasting and post-meal glucose concentrations, some natural compounds ameliorate the glycation process and may help rein in the sugar-induced inflammatory cascade. Chief among these anti-glycation nutrients are benfotiamine, a member of the B-vitamin family, and carnosine, an amino acid.

Benfotiamine has been used to target diabetic complications since the mid 1990's. More recent evidence continues to support its use as a powerful protector against blood sugar-induced tissue damage. In a clinical trial, 165 subjects with diabetes were randomised to receive benfotiamine at either 300 or 600 mg per day, or a placebo for 6 weeks. After the intervention period, those taking benfotiamine exhibited improvements in neuropathic pain in a dose-dependent fashion. An animal model found that benfotiamine relieved neuropathic pain by powerfully suppressing inflammation. Moreover, laboratory experiments have shown that, in addition to blocking glycation reactions, benfotiamine may regulate inflammation more directly by modulating COX and LOX enzyme activity.

Carnosine exerts a range of favorable biochemical effects within the body; it powerfully blunts glycation reactions and eases oxidative stress. In addition, several experiments have revealed a marked ability of carnosine to suppress inflammation in various cell types. Unfortunately, carnosine levels decline as much as 63% between ages 10 and 70. Furthermore, in patients with type II diabetes, skeletal muscle carnosine content is markedly lower than in healthy control subjects. When carnosine is administered as a supplement to animals with chemically–induced diabetes, it is able to protect delicate retinal cells from inflammatory complications related to high blood sugar.

THE INFLAMMATORY REFLEX

Survival is impossible without vigilant defence against attack and injury. The innate immune system continuously surveys the body for the presence of invaders. When it encounters an attack, it involuntarily sets in motion a discrete, localised inflammatory response to thwart most pathogenic threats. The magnitude of the inflammatory response is crucial: insufficient responses result in immunodeficiency, which can lead to infection and cancer; excessive responses cause morbidity and mortality in diseases such as rheumatoid arthritis, Crohn's disease, atherosclerosis, diabetes, Alzheimer's disease, multiple sclerosis, and cerebral and myocardial ischaemia. If inflammation spreads into the bloodstream, as occurs in septic shock syndrome, sepsis, meningitis and severe trauma, the inflammatory responses can be more dangerous than the original inciting stimulus. Homeostasis and health are restored when inflammation is limited by anti-inflammatory responses that are redundant, rapid, reversible, localised, adaptive to changes in input and integrated by the nervous system.

The nervous system is composed of sensory systems (which detect the state of the body and organs) and motor systems (which transmit signals to the body and organs). Whereas the somatic motor system controls voluntary movements, the autonomic motor system controls visceral body functions and innervates glands (involuntary). The autonomic nervous system has two principal divisions, the parasympathetic pathway and the sympathetic pathway, which act either in synergy or in opposition to mediate basic physiological responses in real time.

The autonomic system continuously controls heart rate and blood pressure, respiratory rate, gastrointestinal motility, body temperature and other constantly changing, essential life functions. The autonomic nervous system interacts with the primitive brain, including the limbic system (serving important memory functions), brain stem and hypothalamus. Hypothalamic neural output is relayed to sympathetic and parasympathetic nuclei in the brain stem and spinal cord.

Hormonal input also controls the release of pituitary hormones, which in turn regulate basic functions of the endocrine organs. Autonomic nervous functions are normally subconscious, but essential basic autonomic functions can be placed under conscious control from signals originating in the higher brain (cerebral cortex). For example, subjects can be trained through biofeedback to decrease their heart rate by increasing parasympathetic outflow.

Recent insights have identified a basic neural pathway that reflexively monitors and adjusts the inflammatory response. Inflammatory stimuli activate sensory pathways that relay information to the hypothalamus. Like a knee-jerk reflex, in which the stretching of a patellar tendon elicits a rapidly opposing motor action, inflammatory input activates an anti-inflammatory response that is fast and subconscious. This prevents spillage of inflammatory products into the circulation. The nervous system integrates the inflammatory response: it gathers information about invasive events from several local sites, mobilises defences and creates memory to improve chances for survival.

The neural control of acute inflammation is reflexive, directly interconnected and controllable. Special emphasis is placed on cholinergic anti-inflammatory mechanisms that inhibit the activation of macrophages and the release of cytokines. The evidence indicating that stimulation of the vagus nerve, by either electrical or pharmacological means, prevents inflammation and inhibits the release of cytokines that are clinically relevant drug targets for treating inflammatory disease.

INFLAMMATION MEDIATED BY TNF

Tumour-necrosis factor (TNF), a cytokine with a relative molecular mass of 17,000 (M_r 17K), is produced by activated macrophages in response to pathogens and other injurious stimuli, and is a necessary and sufficient mediator of local and systemic inflammation. Local increases in TNF cause the cardinal clinical signs of inflammation, including heat, swelling, pain and redness. Systemic increases in TNF mediate tissue injury by depressing cardiac output, inducing microvascular thrombosis and mediating systemic capillary leakage syndrome. TNF amplifies and prolongs the inflammatory response by activating other cells to release both cytokines such as interleukin 1 (IL-1) and high mobility group B1 (HMGB1), and mediators such as eicosanoids, nitric oxide and reactive oxygen species, which promote further inflammation and tissue injury. TNF is essential for the complete expression of inflammation during invasion, and self-limited inflammation is normally characterised by decreasing TNF activity.

Low amounts of TNF can contribute to host defence by limiting the spread of pathogenic organisms into the circulation, promoting coagulation to localise the invader, and stimulating the growth of damaged tissues. In a typical 'successful' inflammatory response, the duration and magnitude of TNF release is limited, its beneficial and protective activities predominate, and it is not released systemically. Studies of the inflammatory action of TNF in non-malignant disease have led to widespread investigation of both the 'normal' mechanisms that regulate inflammation and the therapeutic potential of monoclonal antibodies specific for TNF.

Monoclonal antibodies against TNF

Early studies using monoclonal antibodies against TNF showed that this approach effectively prevents lethal tissue injury during bacterial invasion. Subsequent clinical trials led to the registration of both monoclonal antibodies against TNF, and TNF-binding proteins for treating rheumatoid arthritis and Crohn's disease. Many individuals with these debilitating inflammatory illnesses have enjoyed disease remissions and an improved quality of life. Crippling joint pain has been alleviated in children with rheumatoid arthritis treated with TNF antibodies; some of the youngest patients have even experienced 'catch-up growth' and normalisation of development (U. Andersson, personal communication). These and other clinical successes with TNF monoclonal antibodies have proved that cytokine responses can be manipulated to specific therapeutic advantage for inflammatory disease.

But strategies using TNF antibodies have not been translated successfully into treatments for bacterial invasion or sepsis, for reasons that have been reviewed extensively elsewhere. Most notably, serum concentrations of TNF were undetectable in most of the individuals recruited into clinical sepsis trials, because the study group comprised a heterogeneous population with diverse diseases at varying stages of illness. In early experiments of the use of TNF monoclonal antibodies in bacteraemia, it became clear that TNF is an early mediator of inflammation and that TNF antibodies are ineffective if therapy is initiated after serum TNF has been cleared. Continued interest in understanding the use of TNF antibodies for individuals with sepsis is now focused on identifying a homogenous study population with increased serum TNF for treatment early in the course of illness.

An alternative therapeutic strategy is to target 'late' mediators of lethality that are produced after TNF in the inflammatory pathway. HMGB1 has been implicated as an experimental therapeutic target that is produced relatively late in the course of endotoxaemia. Antibodies specific for HMGB1 confer significant protection against the lethality of endotoxaemia, even when the first antibody doses are administered after the early TNF concentrations have been cleared. Reducing serum concentrations of HMGB1 by administering ethyl pyruvate as late as 24 h after the onset of sepsis rescues animals from lethality, indicating that it may now be possible to develop therapies for sepsis that cover a significantly wider, clinically relevant treatment window.

Other cytokines have been implicated as therapeutic targets in the pathogenesis of inflammatory diseases, and it is likely that future treatment plans will target mediators in addition to TNF. Neural regulation of inflammation primarily on cholinergic inhibition of TNF, evidence indicates that these neural anti-inflammatory mechanisms also inhibit the release of IL-1, IL-18 and HMGB1.

Anti-inflammatory responses normally inhibit inflammation

Highly conserved, counter-regulatory mechanisms normally limit the acute inflammatory response and prevent the spread of inflammatory mediators into the bloodstream. Activated immunologically competent cells release TNF receptor fragments that bind and neutralise its inflammatory and potentially toxic actions. Anti-inflammatory cytokines, such as IL-10 and transforming growth factor- (TGF-), specifically inhibit the release of TNF and other proinflammatory mediators. Adrenal glucocorticoids, adrenaline, -melanocyte-stimulating hormone >-MSH) and other 'classical' stress hormones inhibit cytokine synthesis and intracellular signal transductio>. Spermine accumulates at sites of tissue injury and infection and inhibits macrophage activation and cytokine synthesis.

The importance of these endogenous anti-inflammatory pathways has been shown by experimentally impairing isolated pathways. For example, animals subjected to hypophysectomy or adrenalectomy are significantly sensitised to the lethal effects of endotoxin. In the absence of an adequate adrenocorticotropic hormone (ACTH) and glucocorticoid response, TNF is significantly overexpressed during endotoxaemia. Functional deficiencies in the release of corticotropin-releasing factor (CRF) predispose Lewis rats to developing experimental arthritis induced by streptococcal antigens because of an insufficient glucocorticoid response. Animals deficient in IL-10 develop a chronic inflammatory bowel disease that predominately affects the colon and are susceptible to a more severe form of collagen-induced arthritis.

Administration of specific pharmacological spermine antagonists significantly increases local TNF activity and carrageenan-induced oedema formation, and amplifies the inflammatory response. Together, these findings show that loss of endogenous anti-inflammatory mechanisms converts a normally protective, self-limited inflammatory response into an excessive, potentially deleterious response.

COMMUNICATION BETWEEN IMMUNE AND NERVOUS SYSTEMS

The activation of pituitary-dependent adrenal responses after endotoxin administration provided early evidence that inflammatory stimuli can activate anti-inflammatory signals from the central nervous system (CNS). Subsequently, Besedovsky *et al.* showed directly that inflammation in peripheral tissues alters neuronal signalling in the hypothalamus. Extensive work has identified a common molecular basis for communication, with cells from each system expressing signalling ligands and receptors from the other. For example, neurons in the CNS can synthesise and express TNF and IL-1, and these cytokines may participate in neuronal communication. This communication is bi-directional, because cytokines can activate hypothalamic-pituitary release of glucocorticoids and, in turn, glucocorticoids suppress further cytokine synthesis. In addition, cells of the immune system can produce neuropeptides (including endorphins), acetylcholine and other neurotransmitters.

The importance of the interaction between the nervous system and immune system signalling has been demonstrated recently in the development of pathological pain. Watkins and Maier have proposed that cytokines produced by inflammatory and glial cells change neuronal excitability and that this link contributes directly to the development of intractable pain.

Cholinergic anti-inflammatory pathway

Our understanding of the basic mechanisms that regulate inflammation has been advanced by the identification of a neural mechanism that inhibits macrophage activation through parasympathetic outflow. Called the 'cholinergic anti-inflammatory pathway' because acetylcholine is the principle parasympathetic neurotransmitter, macrophages that are exposed to acetylcholine are effectively deactivated. The vagus nerve (which was named for its wandering course) innervates the principal organs, including those that contain the reticuloendothelial system (liver, lung, spleen, kidneys and gut). Experimental activation of the cholinergic anti-inflammatory pathway by direct electrical stimulation of the efferent vagus nerve inhibits the synthesis of TNF in liver, spleen and heart, and attenuates serum concentrations of TNF during endotoxaemia. Vagotomy significantly exacerbates TNF responses to inflammatory stimuli and sensitises animals to the lethal effects of endotoxin.

This 'hard-wired' connection between the nervous and immune systems functions as an anti-inflammatory mechanism in other models of systemic and local inflammation. Direct stimulation of the vagus nerve *in situ* inhibits proinflammatory cytokine synthesis in liver and cardiac tissue obtained from animals subjected to ischaemia-reperfusion by transient aortic clamping. Stimulation of either the right or the left cervical vagus nerves protects against the development of hypotension and inhibits serum TNF responses to ischaemia reperfusion.

The protection conferred by stimulation of the vagus nerve is dependent on the applied voltage and is associated with normalisation of tachycardia during the reperfusion-induced hypotensive phase. In a standardised model of experimental murine arthritis induced by the application of carrageenan, vagus nerve stimulation inhibits the inflammatory response and suppresses the development of paw swelling, indicating that the cholinergic anti-inflammatory pathway can inhibit localised inflammation specifically.

The molecular dovetail between the cholinergic nervous system and the innate immune system is a nicotinic, -bungarotoxin-sensitive macrophage acetylcholine recepto. Exposure of human macrophages, but not peripheral blood monocytes, to nicotine or acetylcholine inhibits the release of TNF, IL-1 and IL-18 in response to endotoxin. Tissue macrophages, but not circulating monocytes, produce most of the TNF that appears systemically during an excessive inflammatory response. Interaction between the macrophage cholinergic receptor and its ligand inhibits the synthesis of proinflammatory cytokines (TNF, IL-1 and IL-18) but not anti-inflammatory cytokines. Acetylcholine inhibits the expression of TNF protein in macrophages, but not the induction of TNF messenger RNA levels, indicating that activation of the cholinergic receptor transduces intracellular signals that inhibit cytokine synthesis at a post-transcriptional stage.

As compared with macrophages, monocytes are refractory to the cytokine-inhibiting effects of acetylcholine: only supraphysiological concentrations of cholinergic agonists inhibit cytokine synthesis in monocytes. The macrophage acetylcholine receptor is distinct from the muscarinic receptor activities identified on lymphocytes, peripheral blood mononuclear cells and alveolar macrophages. The exquisite sensitivity of macrophages to acetylcholine suggests that other non-neuronal cells that produce acetylcholine (such as epithelial cells, T lymphocytes and endothelial cells) might also participate in modulating the function of adjacent tissue macrophages.

Vagus nerve stimulation suppresses inflammation

Stimulation of efferent vagus nerve activity has been associated classically with slowing heart rate, induction of gastric motility, dilation of arterioles and constriction of pupils. Inhibition of the inflammatory response can now be added to this list. From an oversimplified, teleological engineering perspective, there are many reasons why a neural-based anti-inflammatory pathway is advantageous. The diffusible anti-inflammatory network, which includes glucocorticoids, anti-inflammatory cytokines and other humoral mediators, is slow, distributed, non-integrated and dependent on concentration gradients. By contrast, the cholinergic anti-inflammatory pathway is discrete and localised in tissues where invasion and injury typically originate.

Neural regulation of discrete, distributed, localised inflammatory sites provides a mechanism for integrating responses in real time. It is intriguing to consider that, in addition to the development of immunological memory, the involvement of the cholinergic anti-inflammatory pathway might also modulate processing events that promote neural memory of the peri-inflammatory events (that is, the 'hissing snake' or 'charging lion' that caused the wound and/or infection). Clark *et al.* recently discovered that electrical stimulation of the vagus nerve in humans significantly enhanced word-recognition memory, indicating that memory formation and vagus nerve activity are closely linked.

SENSORY FUNCTION OF VAGUS NERVE SIGNALS IN INFLAMMATION

The CNS receives sensory input from the immune system through both humoral and neural routes. Blalock originally suggested that the immune system functions as a 'sixth sense' that detects microbial invasion and produces molecules that relay this information to the brain. TNF and other immunological mediators can gain access to brain centres that are devoid of a blood–brain barrier in the circumventricular region. Indeed, the dorsal vagal complex, comprising the sensory nuclei of the solitary tract, the area postrema and the dorsal motor nucleus of the vagus, responds to increased circulating amounts of TNF by altering motor activity in the vagus nerve. This humoral route for communication between the immune system and the nervous system has been implicated in the development of fever, anorexia, activation of hypothalamic-pituitary responses to infection and injury, and other behavioural manifestations of illness.

Sensory innervation of immune organs by ascending fibres travelling in the vagus nerve, as well as by other pain and ascending sensory pathways, provides important input about the status of invasive and injurious challenges in distributed body compartments. Notably, these neural inflammation-sensing pathways can function at low thresholds of detection and can activate responses even when the inflammatory agents are present in tissues in quantities that are not high enough to reach the brain through the bloodstream. Watkins and colleagues have provided insight into the sensory role of afferent vagus nerve fibres by observing that vagotomy blunts the development of fever in animals exposed to intra-abdominal IL-1. The afferent vagus pathway is activated by very low doses of endotoxin or IL-1; but higher doses of these agents can directly activate thermogenic responses through the humoral route to the brain. It is not completely clear how the vagus nerve 'detects' the presence of low doses of endotoxin or other inflammatory agents, but neurons in the vagus nerve express IL-1 receptor mRNA and discrete IL-1-binding sites have been identified on glomus cells in the vagus nerve proper.

Electrophysiological studies indicate that vagus nerve signals also can be activated by TNF, other cytokines, mechanoreceptors, chemoreceptors, temperature sensors and osmolarity sensors that might be activated at an inflammatory locus. Somatic sensory input into the CNS is organised somatotopically, such that sensory input from a discrete peripheral site is localised precisely in the ascending fibre pathways and brain. The first CNS synapse for afferent vagus signals lies in the nucleus tractus solitarius, and electrolytic lesioning of this region impairs the development of IL-1-induced fever. Thus, inflammation-derived sensory input can be processed differentially in the brain, depending on the location of the inflammatory site and the nature of the sensory signal.

Reflex inhibition of inflammation

The inflammation-sensing and inflammation-suppressing functions outlined above provide the principal components of the inflammatory reflex. The appearance of pathogenic organisms in a local wound, or at the site of epithelial barrier dysfunction, activates innate immune cells that release cytokines. These activate sensory fibres that ascend in the vagus nerve to synapse in the nucleus tractus solitarius. Increased efferent signals in the vagus nerve suppress peripheral cytokine release through macrophage nicotinic receptors and the cholinergic anti-inflammatory pathway. The 'inflammatory reflex' is described as localised, rapid and discrete; but it can also induce systemic humoral anti-inflammatory responses.

This occurs because vagus nerve activity can be relayed to the medullary reticular formation, to the locus ceruleus and to the hypothalamus, leading to increased release of ACTH from the anterior pituitary. Increased cytokine production in tissues causes pain, providing another mechanism for transferring information from the immune system to the brain. This information can be relayed to other brain centres that influence motor output in the vagus nerve. Pain and stress can activate the flight-or-fight responses, and the resultant increase of adrenaline and noradrenaline also can inhibit macrophage activation and suppress synthesis of TNF and other cytokines.

High sympathetic activity and resultant increases in catecholamines stimulate the -adrenergic-receptor-dependent release of IL-10, a potent anti-inflammatory cytokine, from monocyte>. Thus, the anti-inflammatory effects of the sympathetic and parasympathetic nervous systems seem to be synergistic in this setting.

Classical teaching stresses that actions of the sympathetic and parasympathetic nervous systems are usually in opposition. But in many situations the two systems function synergistically. For example, simultaneous stimulation of both sympathetic and vagus nerves produces a higher increase in cardiac output than does isolated stimulation of either nerve alone. Flight-or-fight activation of sympathetic responses also stimulates increased vagus nerve output. The combined action of these neural systems is significantly anti-inflammatory and is positioned anatomically to constrain local inflammation by preventing spillover of potentially lethal toxins into the circulation through both local (neural) and systemic (humoral) anti-inflammatory mechanisms.

Implications of the inflammatory reflex

Knowledge of the inflammatory reflex and the cholinergic anti-inflammatory pathway is yielding insight into both physiological pathways and therapeutic strategies. For example, it may be possible to activate neural anti-inflammatory mechanisms using small molecules that initiate signals in proximal components of the pathway in the CNS. One such molecule is CNI-1493, a tetravalent guanylhydrazone that was originally described as an inhibitor of macrophage activation and TNF release.

CNI-1493 inhibits TNF synthesis and inflammatory responses in animal models of local and systemic inflammation. It also significantly reduced disease severity in a small clinical trial of severe Crohn's disease and is currently being evaluated in a large phase II trial of Crohn's disease. Unexpectedly, recent evidence has shown that the TNF-suppressing activities of CNI-1493 *in vivo* are dependent on the cholinergic anti-inflammatory pathway, and that CNI-1493 functions as a pharmacological stimulator of the vagus nerve: intracerebral application of small doses of CNI-1493 significantly inhibited peripheral TNF synthesis, and intact vagus nerves were required to prevent increases in serum TNF. The mechanism through which CNI-1493 activates the vagus nerve is unknown, but increased vagus nerve firing has been observed after either intracerebral or intravenous administration of CNI-1493 — an effect that seems to be dependent on specific CNS receptors.

It is likely that other experimental and clinically approved therapeutic agents suppress peripheral inflammation by activating pathways in the CNS. Small doses of -MSH applied intracerebrally inhibited pulmonary myeloperoxidase activity in mice exposed to endotoxi and suppressed the development of intradermal oedema induced by exposure to TNF or IL-1. Specific anti-inflammatory responses have been observed in response to intracerebral application of salicylates, but not dexamethasone. The cardiac anti-arrhythmic drug amiodarone has been identified as an inhibitor of TNF synthesis in monocytes *in vitro*, but it also functions as a potent stimulator of vagus nerve activity. Systemic administration of the non-steroidal anti-inflammatory drugs aspirin, indomethacin and ibuprofen substantially increases vagus nerve activity. Although this vagus nerve response had been studied in the context of increasing gastric acidity and ulcer formation, knowledge of the cholinergic anti-inflammatory pathways raises the possibility that the vagus-nerve-stimulating activity of these agents may also contribute to their anti-inflammatory action. A better understanding of the CNS receptors, pathways and neural mechanisms that activate the vagus nerve to inhibit production of TNF should facilitate development of this pharmacological 'vagus-nerve-stimulating' approach.

Another experimental therapeutic approach is based on direct electrical stimulation of the vagus nerve. So far, more than 10,000 individuals have received implantable vagus nerve stimulators for the treatment of epilepsy. Vagus nerve stimulation in humans with small, pacemaker-like devices is safe, well tolerated and not associated with increased rates of infection. But the immunological effects of this approach have not been reported and, indeed, it will interesting to assess whether stimulating the vagus nerve in humans modulates TNF synthesis and inflammation.

In place of implantable devices, it should be possible to develop pharmacological approaches that target the peripheral macrophage receptor to inhibit TNF synthesis. A precedent for this approach has been already achieved in the clinic, because nicotine administration is significantly efficacious in reducing the severity of ulcerative colitis. Other preclinical studies using standard murine models of diabetes have shown that nicotine reduces the incidence of diabetes by reducing pancreatic concentrations of TNF and other cytokines. Unanticipated activities of the cholinergic anti-inflammatory pathway in inflammatory disease and in non-immune cells might be determined by further studies.

Some of the earliest studies of the nervous system and inflammation examined the effects of pavlovian conditioning on intra-abdominal inflammatory responses. Behavioural conditioning using models of learned association can reproducibly influence acute inflammatory responses and alter the course of experimental inflammatory diseases in animals and humans. Hypnosis and meditation can significantly increase vagus nerve output and have been observed to inhibit immediate-type and delayed-type hypersensitivity responses. Biofeedback and acupuncture have been used to modulate vagus nerve activity to alter bowel function, gastric acidity and heart rate.

Each of these approaches has been used to reduce experimental inflammation, but the relationships between vagus nerve activity and anti-inflammatory action had not been defined previously. Autonomic dysfunction occurs as a classical complication of rheumatoid arthritis, diabetes and other autoimmune disorders. It is now intriguing to consider whether vagus nerve dysfunction underlies the progression of inflammation, owing to impairment of the cholinergic anti-inflammatory pathway. It is reasonable to propose that, one day, the rational modulation of vagus nerve activity using these or other approaches may provide a therapeutic advantage for inflammatory disease.

INFLAMMATORY BOWEL DISEASE

Inflammatory bowel disease (IBD) refers to the condition that results when cells involved in inflammation and immune response are called into the lining of the GI tract. This infiltration thickens the bowel lining and interferes with absorption and motility (the ability of the bowel to contract and move food). With abnormal ability to contract and abnormal ability to absorb, the bowel's function is disrupted. Chronic vomiting results if the infiltration is in the stomach or higher areas of the small intestine. A watery diarrhea with weight loss results if the infiltration is in the lower small intestine. A mucous diarrhea with fresh blood (colitis) results if the infiltration occurs in the large intestine. Of course, the entire tract from top to bottom may be involved. Many people confuse inflammatory bowel disease with irritable bowel syndrome, a stress-related diarrhea problem. Treatment for irritable bowel is aimed at stress; it is a completely different condition from IBD.

Infiltration of the bowel with inflammatory cells occurs when something inflammatory (or, in other words, stimulating to the immune system) is happening within the intestinal tract. In food allergy, the digested food stimulates the immune system and causes infiltration of the bowel lining with inflammatory cells. With intestinal parasites, the parasites themselves stimulate the immune system. The World Small Animal Veterinary Association defines IBD as an inflammatory infiltration for which no specific cause can be found. The actual stimulation may be from products released by the bacteria that live in the bowel, from digested food, or something yet unknown. At this point, therapy is directed at suppressing the immunological/inflammatory infiltration.

VETERINARIAN THINK MY PET MIGHT

A little vomiting or diarrhea here and there seems to be pretty standard for pet dogs and cats. After all, cats groom themselves and get hairballs. Dogs eat all sorts of ridiculous things they aren't supposed to. Still, many owners notice that their pets seem to have vomiting or diarrhea a bit more often than it seems they should.

It might be subtle where one notices that one is cleaning up a hairball or vomit pile rather more frequently than with previous pets or it could be the realisation that one has not seen the pet have a normal stool in weeks or months. Typically, the animal doesn't seem obviously sick. Maybe there has been weight loss over time but nothing acute.

There is simply a chronic problem with vomiting, diarrhea or both. IBD is probably the most common cause of chronic intestinal clinical signs and would be the likely condition to pursue first.

Inflammatory Bowel Disease Diagnosed

The diagnosis of IBD requires a tissue biopsy, which is obviously invasively collected with some expense. Since there are a number of other conditions that cause similar signs, a step-by-step testing sequence precedes biopsy.

The first step in pursuing any chronic problem is a metabolic database. This means running a basic blood panel and urinalysis to rule out biochemically widespread problems, such as liver disease or kidney disease, pancreatitis, or hyperthyroidism in cats that could be responsible for the signs. Since IBD is localised to the GI tract, such a database is usually normal but might express a general inflammatory response in the blood or a loss of blood proteins as often there is a leak of albumin, an important blood protein, from the intestine into the bowel contents.

The database not only serves to rule out metabolic causes for the patient's symptoms but also assesses other areas, potentially turning up unanticipated problems and identifying factors that could change what medications are used.

Fecal testing and broad spectrum deworming is often performed at this time to rule out parasitism as a cause of the chronic inflammation. If the patient is young or has been housed with multiple animals, more obscure parasites may be afoot and often special fecal testing is submitted to the laboratory for PCR testing. Typical organisms screened by this kind of testing include: *Giardia, Cryptosporidium, Salmonella, Tritrichomonas,* and *Clostridium perfringens.*

In dogs, a condition called Addison's disease is able to create chronic waxing and waning intestinal disease, among numerous other possible manifestations. This condition, more correctly termed hypoadrenocorticism, is often referred to as "the Great Imitator" as it can mimic many other diseases besides IBD. This condition revolves around a deficiency in cortisol, a crucial hormone in adaptation to stress.

Treatment is relatively straightforward so it is important not to forget to screen for this condition. This is done with a screening test called baseline cortisol blood level or with a longer test called an ACTH stimulation test, which is a more definitive test that requires an hour or two in the hospital.

In both cats and dogs, a trypsin-like immunoreactivity (TLI) test would be performed to rule out pancreatic exocrine insufficiency, a deficiency of digestive enzymes. This condition is relatively easy to treat but, like Addison's disease, cannot be diagnosed without a specific confirming test. Typically this test is run in combination with a vitamin B-12 level and a folate level. When intestinal bacterial populations alter (we used to say "overgrow" but that is not technically accurate), folate levels rise and B-12 levels drop. Antibiotics are likely indicated in this situation as well as vitamin B12 injections.

Somewhere in the course of this work up, ultrasound of the abdomen is generally recommended.

Ultrasound is able to image and enable sampling of areas within the belly that cannot be accessed by endoscopy. The texture of the liver and pancreas are evaluated and the size of the mesenteric lymph nodes, which serve the bowel, are examined.

The layering of the bowel is also evaluated to see if it is thickened, as is more typical of IBD, or more disrupted as with intestinal cancer such as lymphoma. Since cancer is a consideration for most adult animals with chronic intestinal disease, this kind of imaging is particularly valuable and if any unusual textures or even masses are discovered, they may be needle aspirated for analysis.

If this kind of non-invasive testing is not revealing, then the definitive test for IBD is needed: a biopsy. Tissue samples must be harvested from several areas of the GI tract. This can be done either surgically or via endoscopy.

Endoscopy involves the use of a skinny tubular instrument (an endoscope) which has a tiny fibre optic or video camera at the end. The endoscope is inserted down the throat, into the stomach and into the small intestine where small pinches of tissue are obtained via tiny biting forceps. If the large intestine is to be viewed, a series of enemas is needed prior to the procedure as well as a relatively long fast.

The endoscope is inserted rectally and again tissue samples are harvested. The advantage of this procedure over surgery is that it is not as invasive as surgery. Patients typically go home the same day. Disadvantages are expense (often referral to a specialist is necessary) and the fact that the rest of the abdomen cannot be viewed. Growths that are seen via endoscopy cannot be removed at that time and a second procedure typically must be planned whereas, if surgical exploration is used to obtain the biopsy, any growths can also be excised at that time.

Surgical exploration may also be used to obtain samples. The recovery afterwards is typically a couple of days though some patients bounce back immediately. With surgery, other organs can also be sampled and abnormal sections of tissue can be removed. Surgery tends to be more expensive than endoscopy but this depends on the recovery period. Often these two procedures work out to be of similar expense.

Tissue samples obtained are processed by a laboratory and analysed. The infiltration of inflammatory cells is graded as mild, moderate, or severe and the type of cells involved in the inflammation are identified.

INFLAMMATORY BOWEL DISEASE TREATED

Diet

Recent studies have shown that patients with normal albumin levels and without vitamin B12 deficiencies have a 50:50 chance of responding to diet alone - no drugs needed. What sort of diet? The diets that have shown most consistent success are the hydrolysed protein diets.

Hydrolysed proteins are "predigested" so as to create protein segments that are too small to stimulate the immune system. Further, they typically are made with medium chain fatty acids, which are easier to absorb than the more customary long chain fats, and favorable omega 3 to omega 6 fatty acid ratios. In other words, there is more to these diets than just their predigested proteins, but approximately 50% of patients showed good improvement after approximately one month on a hydrolysed protein diet.

Another approach is the use of the novel protein diets. The idea here is that the patient cannot have an immunological reaction to a protein source it has never experienced. (It takes long-term exposure to a protein before the immune system will respond against it, so a new protein should be safe.) This means using an unusual protein such as rabbit, venison, fish (for dogs) or duck, so long as the patient has not been fed these foods before. Again, it takes about a month to see a good response.

Patients that are sick enough to have a low albumin level or low vitamin B12 level are too sick for an approach this conservative. They will need medication.

Medication

The cornerstone of treatment for IBD is suppressing the inflammation. In milder cases of large intestinal IBD, the immunomodulating properties of metronidazole (Flagyl®) might be adequate for control but usually prednisone or its cousin prednisolone is needed. Prednisone will work on IBD in any area of the intestinal tract. In more severe cases, stronger immune suppression is needed (as with cyclosporine, chlorambucil, or azathioprine). Higher doses are usually used in treatment at first and tapered down after control of symptoms has been gained. Some animals are able to eventually discontinue treatment or only require treatment during flare-ups. Others require some medication at all times. Long-term use of prednisone should be accompanied by appropriate periodic monitoring tests due to the immune suppressive nature of this treatment.

In cases where it is particularly important to spare the patients from the side effects of long-term steroids, a medication called budesonide can be used. This medication is not readily absorbed from the GI tract and serves as a topical treatment for the lining of the intestine.

TRY TREATMENT AND SKIP THE EXPENSIVE DIAGNOSTICS

Possibly. Certainly, with IBD the diagnostic tests tend to be much more costly than the treatment. The problem is making sure there is enough confidence in the diagnosis of IBD that there will be no harm in skipping diagnostics. It is not unusual to take the work-up all the way through ultrasound and making a treatment decision based on the information obtained up to that point. If the patient is stable enough, there is time to change the diet or try medications and see how it goes.

The biggest problem in simply putting the patient on prednisone or prednisolone involves the possibility of intestinal lymphosarcoma, also called lymphoma. This type of cancer produces chronic diarrhea or vomiting just as IBD can. Lymphoma is temporarily responsive to prednisone but better responses can be obtained from stronger chemotherapy agents. Exposure to prednisone will make the lymphoma much more difficult to diagnose should biopsies be obtained later. Plus, exposure to prednisone can lead to resistance to other medications. (This is less of a problem for cats, but in dogs even a few days of prednisone can make a lasting remission impossible to achieve.)

In short, if you try prednisone or prednisolone without confirming a diagnosis, harm can be caused should the pet have lymphoma instead of IBD. Sometimes it is financially impossible to complete the ideal test sequence so it is important to discuss all the pros and cons with your veterinarian if going this route. Inflammatory bowel disease continues to be a common cause of chronic intestinal distress in both humans and animals. Research for less invasive tests and for newer treatments is ongoing.

THE SYSTEMIC REACTION DURING INFLAMMATION

The response to infection and injuries usually involves a large number of changes both local and distant from the site of inflammation. This physiological condition takes place at the very beginning of the inflammatory process and lasts for 1-2 days. After that, the host returns to normal functions. The systemic response can also be prolonged, if acute inflammation becomes too chronic. These events lead to a wide-ranging systemic response that is also called acute-phase.

The purpose of acute-phase reaction is to counteract the underlying challenge in order to restore the homeostasis as soon as possible. This result is accomplished by isolating and destroying the infective organisms, or removing the harmful molecules, and activating the repair process. Acute-phase reaction includes a wide range of neuroendocrine, hematopoietic, metabolic and hepatic changes.

One of the most interesting features of the acute-phase is the change in the concentrations of many plasma proteins, known as the acute-phase proteins. An acute-phase protein (APP) has been defined as one whose plasma concentration increases (positive acute-phase proteins) or decreases (negative acute-phaseproteins) by at least 25 percent during inflammatory disorders. The very first APP to be described, C-reactive protein (CRP), was discovered in 1930 in the plasma of patients during the acute phase of pneumococcal infection. In some pathologies, CRP may increase more than 1000 times.

Currently, approximately 40 proteins are considered APP. The following review will focus on the knowledge that has been acquired on acute-phase reaction, in particular acute phase proteins in human and domestic animals.

THE REGULATION OF ACUTE-PHASE CHANGES BY CYTOKINES

The synthesis and release of plasma APP from the liver is regulated by inflammatory mediators. These mediators fall into four major categories: IL-6-type cytokines, IL-1-type cytokines, glucocorticoids, and growth factors.

Cytokines mainly stimulate the APP gene-expression, while glucocorticoids and growth factors modulate cytokine activity. Binding of the inflammatory mediators to their respective receptors on hepatocytes and the transduction of this signal induce changes in APP gene expression that are primarily regulated at a transcriptional level.

Cytokines are a group of proteins acting as intracellular and intercellular signalling molecules. The role of cytokines during inflammation is both initiation and fine-tuning of the whole process: some cytokines initiate and amplify the response, others sustain or attenuate it, and some of them cause it to resolve.

During inflammation, inflammatory cells, mainly macrophages and neutrophils that assemble at the site of challenge, together with endothelial cells, secrete the so-called pro-inflammatory cytokines TNF-a, IL-1b, and IL-6, in that order. This order is important, since each cytokine fulfills a precise role in up-regulating or down-regulating the expression of the others.

Pro-inflammatory cytokines induce a number of local and systemic responses: a) the expression of selectins on local endothelial cells, that recruit inflammatory cells from the bloodstream; b) the activation of recruited cells in developing their optimal defensive activity, i.e. increased expression of receptors (Complement receptors), or cell metabolism, such as oxidative burst, and c) expression of other cytokines, such as chemokines, that recruit more and more defensive cells to the inflammation site.

The systemic response develops in two waves: , at the local reaction site these cytokines activate cells such as fibroblasts and endothelial cells to initiate the secondary release of cytokines that also induce endocrine effects. This secondary wave and the consequent appearance of these cytokines in the circulation are responsible for the start of a wide range of systemic inflammatory effects.

Cytokines operate in a very complex signalling network that is still not completely understood. For example IL-1b induces the production of IL-6, but IL-6 inhibits the expression of the other pro-inflammatory cytokine, TNF-a, thus introducing a controlling step in the amplification of inflammation.

Circulating IL-6 is believed to play the most important role in the induction of acute phase reactions. The synthesis of IL-6 is induced by other cytokines, including IL-1b and TNF-a, but also directly by bacterial endotoxins.

Pro-inflammatory cytokines act by binding to their respective receptors on the cell surface. Then transmit the signal into the cell. The expression of the different APP is accomplished either by activation of resident pools of inactive factors in the cytoplasm and/or by increased factor biosynthesis. The signaling pathways that mediate the overexpression of most APP. The signal transmitted by IL-6 binding to their receptors results in a phosphorylation of transcription factor NF-IL-6, which translocates in the nucleus and mediates the transcription of APP genes.

Also NF-IL-6b gene, a de novo synthesised nuclear factor that enters the nucleus, contributes to the amplification of APP genes expression. TNF-a binds to TNF-a receptor, which activates a different nuclear factor, the NF-kB. NF-kB is one of the main factors involved in the induction of gene transcription during AP-response. NF-kB resides in the cytoplasm, associated with its inhibitor IkB.

The signal transduction eventually leads to an activation of IkB kinases that phosphorylate IkB. This phosphorylation targets IkB for degradation via the ubiquitin-proteasome pathway. The NF-kB can therefore be released in the cytoplasm, and move into the nucleus, where it promotes the transcription of AP-genes. IL-1b binds to its receptor IL-R, which can activate both NF-IL6 and NF-kB pathways.

The liver is not the only organ able to produce APP: many of them are also produced extrahepatically, e.g. ceruloplasmin (Cp), complement components, and serum amyloid A protein (SAA).

Classification of APP

Several classifications of acute phase proteins have been proposed. All the up-regulated proteins have been called “positive APP”, in order to differentiate them from the so-called “negative APP”, that are down-regulated. The most physiologically expressed protein, albumin, and several other proteins usually present in blood belong to the latter group.

APP that considers the defensive systems that are activated during systemic inflammation. A different classification of APP production depends on the rate of their expression. The apolipoproteins serum amyloid SAA1 and SAA2 (commonly called SAA) are considered “major acute-phase proteins”. Their concentration can increase as much as 1000-fold. Other major APPs are CRP and a1 acid-glycoprotein (AGP). Complement components are modestly induced during acute-phase. Both coagulation and fibrinolytic proteins and transport proteins, together with the proteases inhibitors proteins, are also slighty up-regulated (up to three fold).

Post-translational regulation of APP production

Post- translation mechanisms, APP modelling and export, may also be involved in this process , for example in AGP expression. An alteration of carbohydrate moiety, with increased sialylation, has been described for AGP, as previously mentioned. An increase in fucose content has also been described in the glycosylation pattern of haptoglobin (Hp) in the dog. This post-translation mechanism is very complex and still not well understood, but it is likely very important during inflammatory response. Increases in glycoforms expressing di-antennary glycans, and an increase in the degree of 3-fucosylation are apparent from the very first moment of AP. More interesting, pro-inflammatory cytokines (IL-1b, IL-6, TNF-a) and glucocorticoids seem to be involved in these modifications.

Proteins whose plasma concentrations decrease: Albumin, Transferrin, Transthyretin, *a*2-HS glycoprotein, Alpha-fetoprotein, Thyroxine-binding globulin, Insulin-like growth factor I, Factor XII.

The function of Acute Phase proteins in the defence of the organism

Probably all the acute-phase proteins have the potential to influence one or more stages of inflammation. For some of them the beneficial effects are evident.

The classic Complement system proteins, most of which are APP, have an important pro-inflammatory role in immunity. The mannose-binding lectin, that activates the alternative pathway of the complement, is also overexpressed during an acute phase. The function of complement pathways during inflammation are well known, and include chemotaxis, plasma protein exudation at inflammatory sites, and opsonisation of infectious agents and damaged cells. Inflammation is a very complex and finely orchestrated process involving many cell types and molecules. A number of the participating molecules are multifunctional and contribute to both the initiation and the control of inflammation at different points during its evolution. In the coagulation and the fibrinolytic system, the overall improvement of defence and restoration of tissue integrity is evident. The requirement for wound healing can be satisfied by the up-regulation of several acute-phase proteins: fibrinogen, for example, causes activation of the endothelium, including an increase in cell adhesion, spreading and proliferation. One of the most striking characteristics of the overexpression during acute phase is that the whole coagulatory and wound-healing system, i.e. fibrinogen, plasminogen, tissue plasminogen activator (TPA), urokinase and plasminogen activator inhibitor-I (PAI-1), is overexpressed, in order to avoid the unbalancing of the repair process. Hp, whose main function is to bind hemoglobin and thus prevent loss of iron, is also considered an angiogenic factor.

Inflammation is a very violent process: the defence of homeostasis from internal and external enemies is often accomplished at the cost of wide "collateral damages" , such as large destruction of tissues. This is the reason why several APP may also have anti-inflammatory activity. Hp, hemopexin and ceruloplasmin exhibit an anti-oxidant function focused against reactive oxygen intermediate (ROI) species.

Several proteases are activated during AP, like phagocytic lysosomal proteases, and the clotting cascade proteases. If not controlled, their adverse effect can override their beneficial one. Moreover, bacterial proteases, such as collagenases, are also active in septic AP. Therefore, the overexpression of protease inhibitors is functional to control both the physiological repair mechanisms and bacterial aggressiveness.

Some metal chelating proteins, such as ceruloplasmin, that binds copper, and hemopexin, that binds heme, also accomplishes direct defence against pathogens. As usual, their role is multifunctional. Both reduce the availability of copper and iron for bacterial growth, meanwhile avoiding the loss from the organisms.

Other proteins are directly involved in the innate immunity against pathogens. LPS-BP, for example, interacts with bacterial lipopolisaccharides and transfers them to CD-14, a receptor on the surface of macrophages and B-cells. Following the presentation of LPS by LPS-BP, a lipopolisaccharides recognition complex is formed on the membrane via the recruitment of a second receptor, Toll Like Receptor 4 (TLR4). These events drive the signaling pathway of TLR, that induce the activation of several inflammatory and immune-response genes, including pro-inflammatory cytokines..

For other proteins, such as CRP, SAA and AGP, the biological advantage is less evident and still not precisely defined. These three proteins belong to the "major APP" group.

Pathophysiologic roles of CRP during the inflammatory process are complex and apparently inconsistent. C-reactive protein is a component of the innate immune system. It binds phosphocholine and therefore can recognise some foreign bacteria as pathogens, but also the phospholipid constituents of damaged cell.

CRP can activate the complement system when bound to one of its ligands and can also bind to phagocytic cells thus initiating the elimination of targeted cells. Finally, CRP induces the expression of inflammatory cytokines and tissue factor in monocytes. Interestingly, the net effect of an overexpression of CRP may result in an anti-inflammatory effect , probably due to the down-regulation of the surface expression of L-Selectin. At high concentration CRP also inhibits the generation of superoxide in neutrophils, and stimulates the synthesis of IL-1R antagonist by monocytes. The function of SAA is not known. No individual lacking the capacity to up-regulate SAA during AP has been described, thus confirming that the protein carries out an essential function. Following AP, SAA associates with HDL_3, replacing Apo-A1 as the predominant apoprotein on this class of HDL.

SAA associated HDL_3 seems to facilitate the uptake and removal of cholesterol from monocytes/macrophages at the inflammatory site. Lindhort *et al* demonstrated that mouse SAA levels during inflammation peak at the same time as plasma cholesterol levels. Moreover, cholesterol directly injected into an experimental abscess is redirected to the plasma: therefore it has been suggested that SAA reversibly controls cholesterol transport to optimise its clearance from dead cells at inflamed sites.

Several other SAA activities have been described: it increases the cleaving of triacylglicerols into glycerol and fatty acids on HDL_3, by enhancing the activity of secretory phospholipase. SAA directly acts on cholesterol molecule, decreasing its esterification and increasing its uptake by hepatocytes.

Free SAA, but not HDL_3 associated, mediates chemotaxis of monocytes, granulocytes and T-cells. Also, SAA exhibits also some inhibitory effects on inflammation , including down-regulation of fever, pgE_2 synthesis, platelet activation and oxidative burst.

The very high expression rate of SAA raises a completely different pathological problem in the organism: the continuous high expression of SAA is the prerequisite for the development of secondary amyloidosis, caused by the conformational change of SAA in an insoluble proteolytic peptide, AA, that deposits as insoluble plaque in major organs. Moreover, since SAA is also synthesised by cells involved in inflammation (monocytes and endothelial cells), and is closely associated with platelets and cholesterol, the possibility that SAA is involved in the pathogenesis of atherosclerosis cannot be ruled out,.

a-1-acid glycoprotein is usually considered one of the major APPs, even if the concentration rises only 2-5 fold. AGP belongs to the lipocalin family, a group of proteins sharing a similar three-dimensional structure capable of binding and carrying hydrophobic molecules.

AGP is a very unusual protein indeed. Its pI is very low and the carbohydrate content is very high (45%). AGP is considered a natural anti-inflammatory and immunomodulatory agent due to its anti-neutrophil and anti-complement activity. AGP inhibits neutrophil activation and increases the secretion of IL-1R antagonist by macrophages..

One of the most interesting characteristics of AGP is that its immunomodulatory activity has been shown to depend on its glycosylation. Inflammation induces not only an overexpression of AGP, but also a modification of its carbohydrate moiety, the sialyl-Lewis X being the most represented. The sialyl-Lewis X form of AGP induced during inflammation reduces both complement and neutrophil mediate injuries , while the non-sialyl Lewis X form does not. Sialyl Lewis X is the ligand for the cell adhesion molecules E- and P-selectin involved in the rolling of leukocytes to endothelial cells and platelets. Since it has been demonstrated that AGP expressed during inflammation actually bind an E-selectin-IgG chimeric molecule, it has been hypothesised that the inflammation-induced increase in sialyl-Lewis X glycans on AGP can be considered a mechanism for feedback inhibition of leukocytes extravasation.

AGP exhibits also a potent platelet aggregation inhibitory activity. Finally, inhibition of oxidative metabolism has also been reported. For all these activities, it may be suggested that the main function of AGP consists in reducing the cellular damage that usually happens as a consequence of the inflammation process.

Why negative APP?

There are no clear reasons to explain the down-regulation of some proteins. It is possible to speculate on the need to divert available amino acids to the production of other acute-phase proteins during systemic response. The amino acids necessary for APP synthesis derive in part from reduced synthesis of other proteins that are not considered important in that very defensive moment, since the decreased production of negative acute phase plasma proteins is not important for host defence, and in part from muscle protein degradation via the ubiquitin-pathway. For what concerns the other down-regulated proteins, has been demonstrated that transthyretin exhibits inhibitory activity of IL-1b production by monocytes and endothelial cells and therefore a decrease in its plasma concentration may be considered as a proinflammatory mechanism.

Other systemic effects of AP phenomena

Apart from the hepatic APP overexpression, the other two main AP effects are fever and leukocytosis. Several cytokines may induce fever, but the final step is driven by IL-6 produced in the brain stem. Esogen pyrogens, such as LPS, activate macrophages and other leukocytes to release endogen pyrogens (the pro-inflammatory) which stimulate the anterior hypothalamus to produce prostaglandins which lead to sympathetic nerve stimulation, vasoconstriction of skin vessels and increase in body temperature, fever. Fever is beneficial up to a certain point. It can inhibit the growth of some microorganisms and increase the rate of enzyme reactions, thus speeding up the body's metabolism.

An increase in the rate of metabolism can increase the rate of phagocytosis, immune responses, and tissue repair. Too high a body temperature, however, may cause serious damage. Leukocytosis can be defined as an increased concentration of white blood cells. The phenomenon of leukocytosis is essential in the primary host defence against infections. The leukocytosis during AP is essentially induced by IL-1b and TNF-a, which mobilise hematopoietic cells egress into the peripheral circulation. IL-1b up-regulates also GM-CSF gene, indispensable for the growth and development of granulocyte and macrophage progenitor cells. Moreover, it stimulates myeloblasts and monoblasts and triggers irreversible differentiation of these cells.

DOWN-REGULATION AND CONTROL OF APR AND PATHOLOGICAL CONSEQUENCE OF ITS FAILURE

Optimal APR should begin after several minutes, the first APP being detected within one-two hours, and last for 1-2 days. Afterwards, the host should return to normal function. The down-regulation of APR involves many inflammatory mediators, such as glucocorticoids, cytokines including IL-4 and IL-10, and receptor antagonists for certain pro-inflammatory cytokines, such as IL-1Ra and IL-6Ra. Glucocorticoids (cortisol) play a major role in modulating the APR. Cortisol enhances IL-6-mediated APP production. It also reduces the release of pro-inflammatory cytokines, decreases capillary permeability and leucocyte recruitment, stabilises lysosomal membranes, and suppresses cells of the immune system. If acute inflammation becomes too chronic the acute-phase reaction can also be prolonged.

APR is not uniformly beneficial. A failure to control acute-phase process has severe pathological consequences. Secondary amyloidosis due to elevated SAA concentration in patients with chronic inflammatory conditions has been previously mentioned. Inflammation-associated cytokines have been implicated in the pathogenesis of anemia in chronic diseases; examples of their involvement include the decreased responsiveness of erythrocyte precursors to erythropoietin, decreased production of erythropoietin, and impaired mobilisation of iron from macrophages.

Septic shock is the consequence of an uncontrolled activation of host response against bacteria. Very aggressive stimulation, or an uncontrolled AP systemic response, inevitably results in an over-activation of pro-inflammatory cytokines that directly affect organ function, but also acts indirectly through secondary mediators.These secondary mediators include nitric oxide, thromboxanes, leukotrienes, platelet-activating factor, prostaglandins, and complement. Both these primary and secondary mediators cause the activation of the coagulation cascade, the complement cascade and the production of prostaglandins and leukotrienes. Endothelial cell damage occurs, which affects profusion of the organs and can lead to multiple organ system failure. Activation of the coagulation cascade can then lead to disseminate intravacular coagulopathy (DIC) and adult respiratory distress syndrome (ARDS). Leukocytosis may also be hazardous to the host, since it may be associated with tissue destruction. In septic shock for example, uncontrolled leukocytosis and intravascular leukocyte activation lead to irreversible damage to endothelium and to organs and tissues.

Finally, the persistence of some systemic acute phase phenomena, such as fever, anorexia, gluconeogenesis and muscle protein catabolism because of long-term stimulation, such as in cancer, uncontrolled sepsis or immunological disorders, eventually lead to metabolic disorders that result in cachexia.

APP IN CLINICAL PATHOLOGY AND PHARMACOLOGY

APP affects protein binding to several drugs, and therefore their distribution in blood and their therapeutic availability for organs and tissues. The most important protein in this respect is probably AGP, which binds basic and neutral drugs, like tamoxifen , trimethoprim and erythromycin. The pharmacokinetic implications are important, since variations in AGP levels can consistently alter the free plasma level of the drug, without affecting its total plasma concentration. Moreover, many enzymes that metabolise drugs are down-regulated during the acute phase. Interestingly, AGP also binds molecules that modulate its expression, such as retinoic acid and cortisol.

APPs are of very valuable diagnostic significance. CRP is currently the most important analyte in humans, and provides diagnostic information on the presence of inflammatory lesions. In some diseases, like rheumatoid arthritis, the CRP values also possess prognostic meaning. SAA concentration is also a marker of inflammation, being more sensitive than CRP for inflammatory diseases , but assays for this protein are not currently available for wide scale use.

APP in domestic animals

APR is characterised by a uniform nature. Nevertheless, expression levels can differ widely from species to species, and some proteins that are considered acute phase proteins for one species, may not be the same in another. Plasma APP families have their representatives in different species. For example CRP is practically negligible in healthy humans but is a major APP in man, dog and pig, whereas it is present in healthy ruminants and its serum concentration is only slightly modified during inflammation. Hp is a major APP in ruminants, but in dogs is a constitutive serum protein, and its increase is very modest during disease. Human Hp is a constitutively secreted plasma protein that exhibits only a moderate increase during APR.

FUNCTIONAL RELATIONSHIP BETWEEN INFLAMMATION AND CANCER

The functional relationship between inflammation and cancer is not new. In 1863, Virchow hypothesised that the origin of cancer was at sites of chronic inflammation, in part based on his hypothesis that some classes of irritants, together with the tissue injury and ensuing inflammation they cause, enhance cell proliferation. Although it is now clear that proliferation of cells alone does not cause cancer, sustained cell proliferation in an environment rich in inflammatory cells, growth factors, activated stroma, and DNA-damage-promoting agents, certainly potentiates and/or promotes neoplastic risk. During tissue injury associated with wounding, cell proliferation is enhanced while the tissue regenerates; proliferation and inflammation subside after the assaulting agent is removed or the repair completed. In contrast, proliferating cells that sustain DNA damage and/or mutagenic assault (for example, initiated cells) continue to proliferate in microenvironments rich in inflammatory cells and growth/survival factors that support their growth. In a sense, tumours act as wounds that fail to heal.

Today, the causal relationship between inflammation, innate immunity and cancer is more widely accepted; however, many of the molecular and cellular mechanisms mediating this relationship remain unresolved — these are the focus of this review. Furthermore, tumour cells may usurp key mechanisms by which inflammation interfaces with cancers, to further their colonisation of the host. Although the acquired immune response to cancer is intimately related to the inflammatory response, this topic is beyond the scope of this article, but readers are referred to several excellent reviews.

UNDERSTAND THE ROLE OF INFLAMMATION IN THE EVOLUTION OF CANCER

To understand the role of inflammation in the evolution of cancer, it is important to understand what inflammation is and how it contributes to physiological and pathological processes such as wound healing and infection. In response to tissue injury, a multifactorial network of chemical signals initiate and maintain a host response designed to 'heal' the afflicted tissue. This involves activation and directed migration of leukocytes (neutrophils, monocytes and eosinophils) from the venous system to sites of damage , and tissue mast cells also have a significant role. For neutrophils, a four-step mechanism is believed to coordinate recruitment of these inflammatory cells to sites of tissue injury and to the provisional extracellular matrix (ECM) that forms a scaffolding upon which fibroblast and endothelial cells proliferate and migrate, thus providing a nidus for reconstitution of the normal microenvironment.

These steps involve: activation of members of the selectin family of adhesion molecules (L- P-, and E-selectin) that facilitate rolling along the vascular endothelium; triggering of signals that activate and upregulate leukocyte integrins mediated by cytokines and leukocyte-activating molecules; immobilisation of neutrophils on the surface of the vascular endothelium by means of tight adhesion through $_{41}$ and $_{47}$ integrins binding to endothelial vascular cell-adhesion molecule-1 and MadCAM-1, respectively; and transmigration through the endothelium to sites of injury, presumably facilitated by extracellular proteases, such as matrix metalloproteinases (MMPs).

A family of chemotactic cytokines, named chemokines, which possess a relatively high degree of specificity for chemoattraction of specific leukocyte populations, recruits downstream effector cells and dictates the natural evolution of the inflammatory response. The profile of cytokine/chemokines persisting at an inflammatory site is important in the development of chronic disease.

The pro-inflammatory cytokine TNF- (tumour necrosis factor-) controls inflammatory cell populations as well as mediating many of the other aspects of the inflammatory process. TGF-1 is also important, both positively and negatively influencing the processes of inflammation and repair. The key concept is that normal inflammation — for example, inflammation associated with wound healing — is usually self-limiting; however, dysregulation of any of the converging factors can lead to abnormalities and ultimately, pathogenesis — this seems to be the case during neoplastic progression.

Neutrophils (and sometimes eosinophils) are the first recruited effectors of the acute inflammatory response. Monocytes, which differentiate into macrophages in tissues, are next to migrate to the site of tissue injury, guided by chemotactic factors.

Once activated, macrophages are the main source of growth factors and cytokines, which profoundly affect endothelial, epithelial and mesenchymal cells in the local microenvironment. Mast cells are also important in acute inflammation owing to their release of stored and newly synthesised inflammatory mediators, such as histamine, cytokines and proteases complexed to highly sulphated proteoglycans, as well as lipid mediators.

INFLAMMATION AND NEOPLASTIC SEQUENCE

Peyton Rous was the first to recognise that cancers develop from "subthreshold neoplastic states" caused by viral or chemical carcinogens that induce somatic changes.

These states, now known as 'initiation', involve DNA alterations, are irreversible and can persist in otherwise normal tissue indefinitely until the occurrence of a second type of stimulation (now referred to as 'promotion'). Promotion can result from exposure of initiated cells to chemical irritants, such as phorbol esters, factors released at the site of wounding, partial organ resection, hormones or chronic irritation and inflammation.

Functionally, many promoters, whether directly or indirectly, induce cell proliferation, recruit inflammatory cells, increase production of reactive oxygen species leading to oxidative DNA damage, and reduce DNA repair. Subversion of cell death and/or repair programmes occurs in chronically inflamed tissues, thus resulting in DNA replication and proliferation of cells that have lost normal growth control.

Normal inflammation is self-limiting, because the production of anti-inflammatory cytokines follows the pro-inflammatory cytokines closely. However, chronic inflammation seems to be due to persistence of the initiating factors or a failure of mechanisms required for resolving the inflammatory response. Why does the inflammatory response to tumours persist?

CELLS IN AUTOIMMUNE DISEASE

In imperial times, the Great Wall of China was easily breached and was not in itself a very effective defence against resolute adversaries. Rather, it was a communication route and housed, far from the imperial centre, a string of lonely guards who quickly engaged invaders and slowed their progress, while alerting and beckoning more substantial back-up forces.

Mast cells, which are scattered in skin and mucosa, have been considered in a similar outward-looking perspective. They are the lead effector cells in the immediate responses that can occur when sensitised individuals contact allergen through outer body surfaces.

On a more beneficial note, their importance in early responses to bacterial or parasitic pathogens has become recognised in recent years. In both situations, mast cells also follow up by recruiting larger cohorts of neutrophils and lymphocytes.

Recent studies suggest, however, that this picture may be incomplete and illustrate how mast cells are important in the complex cellular chains that lead to autoimmune disease.

EHRLICH'S "GORGED CELLS"

Mast cells, whose differentiation pathways and heterogeneity are still poorly understood, originate from precursors of the haematopoietic lineage and circulate in blood and the lymphatic system before homing to tissues and acquiring their final effector characteristics. The expansion, homing and maturation of mast cell precursors are influenced by several cytokines including interleukin 4 (IL-4), IL-9 and nerve growth factor (NGF), but stem-cell factor (SCF) binding to its receptor c-Kit seems to be the main drive for their differentiation and survival: SCF-deficient (Sl/Sl^{d}) and c-Kit-deficient (W/W^{v}) mice are largely, albeit not completely, devoid of mast cells.

Mast cell produce an impressively broad array of mediators and cell–cell signalling molecules, and it may be this very breadth that confers on the mast cell its individuality in the immune system. Many of these mediators, including histamine, numerous specific proteases (members of the tryptase and chymase families) and tumour-necrosis factor- (TNF-), are released by triggered exocytosis from rich intracellular stores. The fast release of TNF- is noteworthy because of the pleitropic pro-inflammatory effects of this cytokine, and because mast cell granules are a plentiful source of rapidly mobilisable TNF-, whose usually slower induction is the result of activated synthesis in other cell systems.

On activation, mast cells also rapidly synthesise bioactive metabolites of arachidonic acid, prostaglandins and leukotrienes. A specific programme of gene expression is also activated, leading to *de novo* synthesis of several cytokines (IL-3, IL-4, IL-5, IL-6, IL-10, IL-13, IL-14 and NGF), chemokines (macrophage inflammatory protein 1, monocyte chemoattractant protein 1 (MCP-1) and lymphotactin) and, again, TNF-. This second-wave response comes after the immediate hypersensitivity reactions, which it amplifies. It may also bias the type of secondary events, for example, by moulding the anti-inflammatory T helper 2 (T_H2) bias of T cells in the local response to airway allergens in asthma. Thus, activated mast cells signal to the vascular system through the potent vasoactivity of histamine and arachidonic metabolites, to monocytes and lymphocytes through the chemotactic and differential properties of cytokines and chemokines, and to the connective substratum through the extracellular proteases.

Several triggers can elicit these responses. The best characterised are allergens complexed to immunoglobulin- (IgE) molecules. Because of the unusually high affinity (10^{-10} M) of the Fc receptor (FcR) for IgE (FcR), mast cells are constantly coated with antigen-specific IgE and are, in essence, masquerading as cells of the adaptive immune system. The crosslinking of these surface-bound IgE by antigen leads to activation and degranulation. Other members of the FcR family are also active, in particular the FcRIII receptor. Anaphylatoxins generated by activation of the complement pathway are also potent activators of some mast cells. Bacterial microbes can trigger mast cells through Toll-like receptors (TLRs), endowing them with the broad 'pattern recognition' capability of the TLR system, which is probably an important element of their antibacterial responses. Some cytokines and chemokines activate mast cells, in particular TNF- and MCP-1, which are themselves released by mast cells, thus raising the potential for a positive feedback loop. Finally, activation of mast cells by co-culture with activated T cells has been described, but it is not clear what molecular mediators may be involved. Direct crosstalk by surface molecules on T cells and mast cells may be important in this context.

AUTOIMMUNE DISEASE IN THE BRAIN

The recent spark of interest in a role for mast cells in initiating or propagating autoimmune disease was prompted by studies on multiple sclerosis and its animal model, experimental allergic encephalomyelitis (EAE). Multiple sclerosis is a chronic inflammatory disorder of the central nervous system (CNS), which is characterised by a breach of the blood–brain barrier, mononuclear cell infiltration of white matter and eventual demyelinisation. A similar autoimmune disease can be induced in susceptible rodent strains by injecting different myelin components, including myelin basic protein (MBP), proteolipid protein and myelin oligodendrocyte glycoprotein (MOG).

Both multiple sclerosis and EAE depend critically on pro-inflammatory T helper 1 (T_H1) $CD4^+$ T cells. B cells, and more specifically the antibodies that they produce, may also be important, although this is still under debate. Numerous studies, dating as far back as 100 years, have reported a correlation between the number and/or distribution of mast cells and the development of multiple sclerosis or EAE. Evidence of mast cell activation in the course of the disease came from the demonstration of increasing degranulation and increased amounts of proteolytic enzymes such as tryptase in cerebrospinal fluid. In addition, drugs considered to 'stabilise' mast cells (for example, cromolyn sodium) have been shown to ameliorate the severity of EAE.

Although these observations were highly suggestive of an essential role for mast cells in these CNS autoimmune diseases, the association remained indirect until the recent studies of Brown and colleagues. These researchers showed that mice lacking mast cells (W/W^v mice) develop EAE later and less severely than do control mice in response to injection of MOG. Complementation of W/W^v mice with immature mast cells derived *in vitro* restores typical EAE susceptibility. Mast cell function seems to be the result of binding antibodies, as it was found to be dependent on expression, by the mast cells, of the FcR. Notably, Brown and colleagues subsequently showed that their procedure does not result in reconstitution of mast cells in CNS tissues, suggesting that mast cells might be exerting their crucial influence outside the inflammatory lesion.

Another line of evidence has independently piqued interest in a role for mast cells in multiple sclerosis and EAE. Gene expression profiling of multiple sclerosis brain lesions detected an unexpectedly high contribution of transcripts either derived from mast cells or otherwise associated with the allergic response, including transcripts encoding histamine receptors, proteases and other inflammatory mediators. These findings rekindle interest in the perplexing finding that the transfer of MBP-specific T_H2 cells to healthy recipients unexpectedly provoked a variant form of EAE characterised by eosinophilic infiltrates into the CNS.

AUTOIMMUNE DISEASE IN THE JOINT

A potential role for mast cells in rheumatoid arthritis has also been highlighted recently. Rheumatoid arthritis is a chronic inflammatory disease of the diarthrodial joints. K/BxN mice spontaneously develop a joint disorder that has many similarities to rheumatoid arthritis. Although the development of disease in this model is initiated by T cells, it also requires B cells, and immunoglobulin- (IgG) antibodies from an arthritic donor can induce disease in a healthy host. The target of both the pathogenic T cells and arthritogenic antibodies is the ubiquitous cytoplasmic enzyme glucose-6-phosphate isomerase (GPI). This enzyme and antibodies against it aggregate as immune complexes at the surface of the articular cavity, where they initiate an inflammatory cascade involving the alternative pathway of complement (acting through C5a), FcRs (in particular, FcRIII), neutrophils and cytokines such as IL-1 and TNF-.

Now it seems that mast cells are also important in this disease process. Both Sl/Sl^d and W/W^v mice are resistant to the induction of arthritis by antibodies against GPI. More definitively, reconstitution of these mice with mast cell precursors restores sensitivity to disease induction. Notably, one of the first events detected after injection of arthritogenic antibodies into wild-type mice is mast cell degranulation in the joint but not in other tissues. This very early event is already apparent an hour after antibody administration, before the recruitment of neutrophils. These results prompted the conclusion that mast cells might have an early, coordinating role in this model of rheumatoid arthritis.

The generality of this conclusion is supported by observations from other murine models of rheumatoid arthritis and from individuals affected with rheumatoid arthritis. Mast cells accumulate in the swollen paws of mice suffering from collagen-induced arthritis, and they degranulate during the disease process. Salbutamol is a 2-adrenergic agonist that prevents mast cell degranulation, and this drug had a strong therapeutic effect on the progression of collagen-induced arthritis. Mast cell deficiency was also found to inhibit the course of antigen-induced arthritis in mice, although the effect was rather mild. Mast cells also accumulate in the synovial tissues and fluids of humans suffering from rheumatoid arthritis, reflecting the presence of mast cell chemotactic or survival activities such as SCF and transforming growth factor- in the synovial fluid. The invading mast cells produce several inflammatory mediators, notably TNF-, IL-1 and vascular endothelial growth factor (VEGF). Notably, TNF- can induce further production of SCF by synovial fibroblasts, potentially augmenting mast cell recruitment and thereby creating an amplification loop.

Autoimmune disease in the skin

Bullous pemphigoid seems to present a situation that is highly similar to the one that unfolds in K/BxN mice. This autoimmune skin disease is characterised by subepidermal blisters resulting from auto-antibodies against two hemidesmosomal antigens, BP230 and BP180. The key features of the human disease can be mimicked by injecting neonatal mice intradermally with IgG antibodies directed against murine BP180. The antibody-induced disease has been known for some time to require activation of the complement pathway and the accumulation of neutrophils. Recently, it has been also shown to depend critically on mast cells.

Mast cell degranulation was one of the first responses detected after the injection of antibodies against BP180, occurring only 1 h after administration and preceding neutrophil accumulation and skin blistering. Injection of antibodies against BP180 into mice lacking mast cells (W/W^v or Sl/Sl^d) did not induce bullous pemphigoid, nor did their injection into wild-type mice pre-treated with cromolyn sodium. But mice lacking mast cells that were reconstituted intradermally with mast cells derived *in vitro* showed typical features of disease.

In the absence of mast cells, IgG still accumulated in the skin and the complement pathway was activated to yield C3a and C5a, but neutrophils were no longer recruited to the dermal lesion. Bullous pemphigoid could be induced in mast-cell-deficient mice injected with antibodies against BP180 if neutrophils or the potent neutrophil attractant IL-8 were injected intradermally. Thus, it was concluded that the crucial role of mast cells in murine bullous pemphigoid is to recruit neutrophils to the developing lesion. A similar process might also occur in the human disease, because degranulated mast cells are a prominent feature of the skin blisters of individuals affected with bullous pemphigoid, and mast-cell-derived chemoattractants are present at high concentrations in blister fluids.

There are several other examples of autoimmune disorders in which mast cells have been implicated, although often only by 'guilt by association'. These include Sjogren's syndrome, chronic idiopathic urticaria, thyroid eye disease and experimental vasculitis. For these disorders it will be important to provide evidence, as in the three diseases highlighted here, that mast cells are more than bystanders that become activated in the inflammatory maelstrom and are involved directly in the complex chain of cellular events that lead to autoimmune damage.

The role of mast cells

Where, however, are mast cells positioned in this chain? What triggers them into action, and which are the important relay molecules ? For the antibody-mediated models (pemphigoid and K/BxN arthritis), there is no dearth of candidates that might activate mast cells: the two main consequences of immune complex formation — the production of complement-derived anaphylatoxins and FcR crosslinking — can both trigger mast cells efficiently. It will be important to pinpoint which of these pathways is involved by analysing mast cell degranulation in knockout animals and by reconstituting W/W^v mice with mast cells derived from complement- or FcR-deficient mice.

For the EAE models, in which T cells are classically thought to be the effectors, one might have invoked the effect that activated T cells have on mast cells. But the effectiveness of mast cell reconstitution seems to be dependent on the presence of FcR, pointing to an involvement of antibodies against MOG in this disease. Notably, MOG-induced EAE is the model that is thought to be most dependent on antibodies for lesion development; thus, here again the mast cell contribution may be antibody-dependent. These data do not rule out a direct interaction between T cells and mast cells, and it will be interesting to examine the role of mast cells in 'pure' T-cell-mediated autoimmune diseases, such as diabetes.

The heterogeneity of mast cell populations, their variations in different tissue environments and how they may differentially integrate input from different stimuli are incompletely understood facets of their biology. Is the response of an airway mast cell to an allergen that crosslinks IgE receptors the same as that of a joint mast cell to deposited IgG? Complex interactions take place between the intracellular signals elicited when FcR and FcR are both engaged, and these influence the mediators that are released or induced. It will be important to determine how concomitant triggering of mast cells through the FcR, C5a and other secondary byproducts of immune complexes may be integrated differentially by mast cells, thereby leading to consequences as different as a pemphigus blister or an EAE plaque.

Downstream of mast cell activation, all of the events described in IgE-induced allergic responses have the potential to fan the autoimmune flames. For example, there will be increased permeability of the local vasculature, which will recruit even more immune complexes into the lesion; notably, local oedema is one of the earliest events in the unfolding of antibody-induced arthritis. There will be modifications of vascular adhesive properties contributing to the recruitment of leukocytes by chemokines, comparable to the mast-cell-mediated influx of neutrophils in models of peritonitis.

In the arthritis model, neutrophils are also essential, and it may be that the sequential mast cell/neutrophil tandem will constitute a frequently recurring theme. The very early timing of mast cell degranulation in both the bullous pemphigoid and rheumatoid arthritis mouse models are consistent with that view. In the peritonitis models, TNF- seems to be the essential mediator for neutrophil recruitment. Given the central role that TNF- seems to have in arthritis, it will be interesting to see whether it is also the principal contribution of the mast cell.

In both asthma and arthritis, the worst damage lies not so much in the immediate inflammation as in the subsequent tissue reorganisation and chronic inflammation. Connective tissue proliferation leads to loss of organ function, whether as an eroding pannus in the joint or as thickened and hyperreactive bronchi.

Arthritis, in particular, has been described as a tumour-like anarchic proliferation of synoviocytes. Several mast cell products have strong trophic effects, including classical growth factors (NGF, epidermal growth factor, VEGF), but some of the mast cell proteases also have mitogenic properties. One might propose that mast cells are important contributors in the anarchic joint reconstruction triggered by the autoimmune attack. Last, as suggested by Brown and colleagues, there is the intriguing possibility that mast cell activation also feeds back to the initiating autoimmune responses in lymphocytes.

The release of tissue neo-antigens through proteolysis might contribute to the epitope spreading observed in EAE. Or, as in asthma, the locally released cytokines might bias T-cell phenotypes, enhancing a T_H2 response that would bolster the dangerous production of auto-antibodies.

Autoimmune diseases such as multiple sclerosis or rheumatoid arthritis are complex and involve long and convoluted molecular and cellular chains, with many possible points for therapeutic intervention. Yet the demonstration of an obligate passage through mast cells in these animal models opens the perspective of harnessing agents that modulate mast cell homeostasis or function to treat human disease.

Mast cells have been positioned historically in the private domain of allergists and have been largely ignored by the autoimmunity field. This ignorance can no longer be sustained as the demarcation between autoimmunity and allergy becomes fuzzy. This is illustrated by the anaphylactic reactions induced, under certain conditions, by injecting myelin proteins or peptides into mice or individuals with multiple sclerosis. And the view of mast cells as a ring of outward-looking sentinels can no longer hold. Their scope clearly includes the inner realm as well.

INFLAMMATION AND THERAPEUTIC VACCINATION IN CNS DISEASES

Inflammation of the central nervous system (CNS) may be the result of both innate and adaptive immune responses. In Alzheimer's disease (AD) an innate immune response is triggered by local production of amyloid -protein (>), whereas in multiple sclerosis (MS) an adaptive immune response directed against myelin components initiates inflammation in the CNS. Adaptive immune responses involving antibody- or cell-mediated responses have differential effects in AD and MS, and in animal models of these diseases. In addition, the recent appearance of encephalitis in individuals with AD that have been immunised with A has parallels to underlying mechanisms of cell-mediated adaptive immune responses in MS, in which pro-inflammatory T-cell responses seem to drive the disease. These features of inflammation, which are outlined for AD and MS, are reviewed here in terms of both disease pathogenesis and therapy.

MULTIPLE SCLEROSIS

Multiple sclerosis is an inflammatory disease of the central nervous system characterised by perivascular cuffs of mononuclear cells that include both lymphocytes and macrophages. This infiltration leads to damage of the myelin sheath and the underlying axon. Activation of microglia and astrocytes occurs in MS, but it is secondary to infiltrating lymphocytes. In the initial stages of the disease, the inflammation that occurs in MS is episodic and associated with discrete attacks of neurological dysfunction followed by recovery, which may leave residual neurological damage. Subsequently the disease often becomes more progressive, developing to a stage where there is less inflammation and nervous system damage is caused by a degenerative process initiated by the inflammation. The episodic inflammation that is classic of MS is clearly visualised by magnetic resonance imaging (MRI) scans of the brain after administration of the contrast material gadolinium. Gadolinium crosses an open blood–brain barrier created by the inflammation and highlights discrete areas of inflammation. The duration of enhanced inflammation in individuals receiving weekly MRI scans is 4–8 weeks, and virtually all new lesions enhance in their earliest phases. When the acute inflammation resolves, it leaves a scar and tissue damage.

This can be seen in the three-dimensional MRI images, which were recorded over a 1-yr period in a single individual affected with MS. The new inflammatory focus can be seen appearing adjacent to the ventricle and then beginning to resolve. The inflammatory process of MS is associated with a complex cascade of inflammatory molecules and mediators, including chemokines, adhesion molecules associated with activated endothelial cell walls and matrix metalloproteases.

The cause of the recurrent inflammation in MS is now generally accepted to be autoimmune in nature, that is, a cell-mediated autoimmune attack against the white matter sheath. An alternative explanation for the episodic and chronic inflammation that is the hallmark of MS is the presence of a virus or infectious agent that has persistently infected the nervous system. But although infectious agents have been extensively sought in MS, none has been isolated. Viruses and infectious agents are, however, thought to be important in triggering the immune system and the immune attack on the nervous system. Given the inflammatory nature of the pathological process and the autoimmune hypothesis, one might expect that anti-inflammatory immunosuppressive drugs would reduce inflammation, as measured by MRI imaging, and positively affect the clinical course. Indeed, this has been shown clearly with agents such as mitoxanthrone, a chemotherapy drug, and cyclophosphamide, a chemotherapy drug that is also used in other inflammatory conditions such as lupus nephritis and inflammatory muscle disease. The most widely used drugs in MS, -interferon and glatiramer acetate, have anti-inflammatory and immunomodulatory effects.

ADAPTIVE CELL-MEDIATED IMMUNE RESPONSES IN MS

The adaptive immune system can be classified broadly into cellular and humoral (antibody)-type responses. Among cellular responses, different types or classes of cellular immune responses have been identified that are essential to understanding the mechanisms of the inflammatory process in MS and to devising strategies to control it. The different classes of cell-mediated immune response have important implications for attempts to develop a vaccination strategy not only for MS but also for AD. Cellular immune responses can be classified as T_H1-type or T_H2-type responses, depending on how they differentiate from T_H0 precursors. T_H1 (or pro-inflammatory) responses are induced when T cells differentiate in the presence of interleukin 12 (IL-12), and T_H1 cells are characterised by the secretion of interferon- (IFN-) and inflammatory mediators such as tumour-necrosis factor- (TNF-). T_H1-type responses are important in fighting viral infections, and MS seems to be a cell-mediated autoimmune disease of a T_H1 type. Anti-inflammatory T-cell responses include both T_H2 responses and T cells that have been classified as 'regulatory cells'. T_H2 responses are induced when T cells differentiate in the presence of IL-4, and T_H2-type cells secrete anti-inflammatory cytokines such as IL-4 and IL-10. T_H2-type responses are important in fighting parasitic infections, and T_H1 and T_H2 responses may cross-regulate each other. Another class of T cell comprises regulatory cells that can downregulate T_H1-type inflammatory processes. Different types of regulatory cell have been described. T_H3 cells act primarily through the secretion of transforming growth factor- (TGF-) and are preferentially induced at mucosal surfaces, T_R1 cells (T regulatory cell 1) act primarily through the secretion of IL-10, and $CD4^+CD25^+$ regulatory cells are T cells that express CD25 (IL-2 receptor) and exert potent regulatory function through cell contact and also through cytokines such as IL-10 and TGF-. If MS is a T_H1-type cell-mediated autoimmune disease, it might be possible to regulate the T_H1 responses by the induction of regulatory cell populations.

The induction of T_H1-type myelin-reactive cells and their migration into the nervous system. It is postulated that T_HP (T precursor) myelin-reactive T cells are induced to differentiate into myelin-reactive T_H1 cells when an antigen that crossreacts with a myelin antigen is presented to a T cell by an antigen-presenting cell in the context of IL-12 and co-stimulatory molecules. It is generally thought that viruses with structures that crossreact with myelin antigens act as crossreactive antigens. T_H1 T cells that react with myelin antigens, such as proteolipid protein (PLP), myelin basic protein (MBP) and myelin oligodendrocyte glycoprotein (MOG), cross the blood–brain barrier where the myelin antigens are represented to the T cell by antigen-presenting cells in the brain (microglia cells), and an inflammatory cascade is triggered with the release of inflammatory mediators that cause damage to the myelin sheath and ultimately the underlying axon.

One of the primary animal models for MS, experimental allergic encephalomyelitis (EAE), is induced by immunising different mouse or rat strains with a myelin autoantigen (such as MBP, PLP or MOG) given in complete Freund's adjuvant, which induces a T_H1-type cell-mediated response against the myelin antigen. In EAE, myelin-reactive T_H1-type $CD4^+$ T cells migrate from the periphery into the CNS, where they also initiate a cascade of immune-mediated damage. In animals, EAE can be induced by the adoptive transfer of T_H1-type $CD4^+$ cells specific for one of the myelin proteins.

The hypothesis that MS is a inflammatory T_H1-type disease is supported by several observations. First, it has been shown directly by the effects of -interferon, the prototypic $>_H1$ cytokine, which when administered to individuals with MS caused clinical exacerbations. Second, individuals affected with MS have a T_H1 bias, as indicated by increased concentrations of IL-12 and IL-18, both of which induce IFN- and increase T_H1-type chemokine receptor expression. Last, IL-12-secreting cells in the peripheral blood are linked to inflammation in the CNS, as measured by gadolinium enhancement on MRI imaging: increased numbers of IL-12-secreting cells in the blood are associated with gadolinium enhancement, and cyclophosphamide decreases the number of IL-12-secreting cells, which is linked to clinical response.

In addition to IL-12, it has been shown recently that osteopontin is important in T_H1 differentiation in autoimmune demyelinating disease. The most widely used immunomodulatory drug in MS, -interferon, seems to have two broad mechanisms of action: it decreases-interferon secretion by cells in the peripheral blood and blocks the migration of T cells across the blood–brain barrie.

Observation of Vaccination

The term 'vaccination' stems from the original observation of Jenner and his use of subcutaneous administration of cowpox to prevent the subsequent development of smallpox. Since then, the term vaccination has acquired a broader meaning. According to current immunological theory, vaccination is no longer restricted to administering infectious agents but applies to manipulating the immune system in a manner that regulates or suppresses inflammatory and even non-inflammatory processes that can cause tissue damage. Thus, one can redefine vaccination as 'the generation or induction of an immune response that is beneficial to the host in halting a pathological process', irrespective of whether that process is immune-mediated, autoimmune or even inflammatory.

Thus, vaccination involves not only the use of the immune system itself to correct or to alter abnormal immune responses that cause damage, but the immune system may be used to affect beneficially pathological processes that are neither autoimmune nor inflammatory. A striking example is represented by reports of the effectiveness of active immunisation with A peptide in adjuvant and the passive administration of antibodies against A to clear amyloid deposits and their surrounding glia and neuronal cytopathology from the brains of transgenic mouse models of AD.

It has also become clear that injury to the nervous system by non-immune mechanisms, such as stroke or trauma, may have a secondary stage associated with inflammation and that immune-based therapies can decrease CNS damage. For example, oral administration of MBP in a rat model of stroke decreases infarct size after middle cerebral artery occlusion and this is associated with increased expression of the anti-inflammatory cytokine TGF- in the nervous system, and nasal administration of myelin oligodendrocyte glycoprotein (MOG) has similar effects in a mouse model of stroke (D. Frenkel and H.W., unpublished results). In an extensive series of studies, Schwartz and co-workers have shown that T-cell autoimmunity against myelin antigens can be beneficial in animal models of central nervous system trauma caused by crush injury of the optic nerve or spinal cord contusion. Thus, an 'inflammatory response' directed against nervous system tissue also has the potential to have a protective or beneficial role.

Antigen-specific vaccination in MS

Antigen-specific modulation of the immune system is presumed to be the most specific and potentially least toxic way in which to manipulate the immune system in disease and represents the classic model of vaccination, that is, the induction of an antigen-specific beneficial immune response. For MS, a T_H1-type cell-mediated disease, the strategy is to induce T_H2 or antigen-specific regulatory cells.

Numerous approaches using antigen-specific therapy have been successful in the murine EAE model and some of these have been tested in individuals with MS. The most successful so far has been the use of glatiramer acetate or copolymer 1, which is now an approved therapy for MS. Glatiramer acetate is a random copolymer of four amino acids that was designed to mimic MBP and thus to induce EAE. It does not have encephalitogenic properties but instead works effectively in what seems to be an antigen-specific manner to suppress EAE by generating regulatory T cells. Although glatiramer acetate has several effects on the immune system, it seems principally to be acting as an altered peptide ligand that induces T_H2- and T_H3-type regulatory cells, which react in the CNS to suppress inflammation.

One of the major conceptual conundrums in designing antigen-specific vaccines for MS relates to the issue of which antigen to administer in MS. There is reactivity to several myelin antigens in MS, both because MS seems to be a syndrome rather than a single disease and because of epitope spreading, in which damage caused by a T cell specific for one myelin antigen induces reactivity to another myelin antigen.

This conundrum seems to have been resolved by the phenomenon of bystander suppression, in which antigen-specific myelin-reactive regulatory cells are induced that secrete anti-inflammatory cytokines such as IL-10 and TGF-. Such regulatory cells secrete anti-inflammatory cytokines when they encounter the autoantigen in the target tissue and thus suppress inflammation in the CNS caused by T cells of a different specificity.

Thus, in the EAE model, one can suppress PLP-induced EAE by glatiramer acetate, by mucosal administration of MBP or by the use of altered peptide ligands of MBP, all of which induce anti-inflammatory regulatory T-cell responses (T_H2, T_H3). Of note, in immune-deficient mice, T_H2-type responses can induce a form of EAE.

But therapeutic vaccination is not without potential risks both in MS and in AD. In the early 1980s, Jonas Salk and colleagues attempted to treat individuals with MS by injecting large amounts of MBP subcutaneously to 'vaccinate' against putative harmful T-cell responses to MBP. They could induce both cellular and humoral (antibody) immune responses to MBP but obtained no consistent positive clinical effects and even some suggestion that the injections might have been harmful.

To obviate harmful sensitisation by injection of MBP, an analogous approach was undertaken using an altered peptide ligand of MBP in which key amino acid sequences had been altered so that injection caused a T_H2 or T_H3 response as opposed to a T_H1 response. Results of a phase II trial showed that injections of large doses of the peptide led to a worsening of MS inflammation in some people, as measured by gadolinium-enhanced lesions on brain MRI, and an increased number of cells reactive to MBP. As part of a larger trial in individuals given a smaller dose, however, positive effects were observed on MRI and immune deviation towards T_H2-type responses was observed.

An A vaccine developed for use in AD has been found to cause adverse effects, which were most probably related to the induction of T_H1-type T-cell responses against A. Of note, T-cell vaccination with myelin-reactive T cells to downregulate pathogenic T_H1 responses has been applied successfully to the EAE model and is being tested in individuals with MS, but it is not applicable to AD because there is no evidence of a pathogenic adaptive T-cell response in AD. DNA vaccination is another approach for treating CNS autoimmune diseases such as MS and has been used effectively in the EAE model by several investigators.

Alzheimer's disease

Alzheimer's disease is the most common form of age-related cognitive failure in humans. It is characterised neuropathologically by the progressive accumulation of the 42-residue A peptide in limbic and association cortices, where some of it precipitates to form a range of amorphous and compacted extracellular plaques. These plaques, particularly the more compacted ones, are associated with dystrophic neurites (altered axons and dendrites), activated microglia and reactive astrocytes. Some of these dystrophic neurites contain intracellular bundles of abnormal paired helical filaments composed of insoluble, hyperphosphorylated forms of the microtubule-associated protein, tau. Paired helical filaments also accumulate in large cytoplasmic masses, called neurofibrillary tangles, in the cell bodies of innumerable limbic and neocortical neurons. The detection of neuritic (amyloid) plaques and neurofibrillary tangles in brain regions important for memory and other cognitive functions provides the basis for confirming a clinical diagnosis of AD after death.

Although it has become increasingly recognised that inflammation may be important in the neuropathological damage that occurs in AD, unlike MS the inflammation in AD seems to arise from inside the CNS with little or no involvement of lymphocytes or monocytes beyond their normal surveillance of the brain.

The inflammatory cytopathology (microgliosis, astrocytosis, complement activation, increased cytokine expression and acute phase protein response) is thought to represent a secondary response to the early accumulation of A in the brain. This innate immune response that occurs in the brain, which is presumably secondary to amyloid deposition, leads to the accumulation of inflammatory mediators such as TNF-, IL-1, IL-6, free radicals and microglia activation.

To what degree this activation of microglia and other potential antigen-presenting CNS cells and secretors of cytokines is involved in the progressive neurodegenerative process is not yet clear, although it has been generally assumed to do more harm than good. Studies of transgenic mice that overexpress an AD-causing mutant form of human amyloid precursor protein (APP) and develop amyloid deposits have shown, however, that crossing such mice with mice overexpressing a natural inhibitor of complement C3 results in a worsening of A plaque load and more neuronal loss. This result suggests that the inflammatory changes found in AD and mouse models thereof, including activation of the classical complement cascade, may represent a beneficial response, at least in part. Nonetheless, clinical studies suggest that conventional anti-inflammatory drugs such as those used in arthritis may delay or slow the progression of AD.

Despite the fact that only local innate inflammation occurs in AD, the theory and immune mechanisms of therapeutic vaccination with reference to MS have unexpectedly become relevant to AD, because the induction of specific adaptive immune responses has been shown to be of benefit in the animal model of AD. It has been discovered that parenteral immunisation of APP transgenic mice with synthetic A in complete Freund's adjuvant can markedly decrease the number and density of A deposits in the brain, with concomitant improvements in neuritic dystrophy and gliosis. Positive effects have also been found after repetitive mucosal (intranasal) administration of the peptide to transgenic mice.

It seems that the induction of antibodies against A has a primary role in the vaccine-mediated clearance of A from the brain, because passive transfer of A antibodies has shown similar beneficial neuropathological effects. Notably, a single parenteral administration of a monoclonal antibody against A has been shown to produce rapid (within hours) benefits on certain behavioural measures of cognitive function in a mouse model, apparently by interfering with some diffusible, putatively synaptotoxic form of A (for example, A oligomers) without lowering the overall amount of A deposits in the brain. Two broad theories about the mechanisms by which A antibodies work in mice have emerged. First, evidence of Fc-mediated uptake and clearance of A antibody complexes by local activated microglia has been obtained. Second, evidence of a net movement of A peptide out of the brain as a result of its binding and mobilisation by A antibodies, both peripherally (in the serum) and centrally (in the cerebrospinal fluid), has been provided. These two proposed mechanisms are not mutually exclusive, and there may be additional ways in which antibodies decrease A-mediated synaptic and neuronal dysfunction. So far there is no clear evidence that T cells have either a protective or an injurious effect in AD or its mouse models, but this possibility needs further research. T-cell responses seem to have a role in the generation of meningoencephalitis after > vaccination to induce antibodies.

HUMAN TRIALS OF A VACCINATION IN AD

The finding that active vaccination with A could profoundly reduce quantities of A peptide in an animal model led to early clinical trials in which an A_{1-42} synthetic peptide was administered parenterally with a previously tested adjuvant (QS21) to individuals with mild to moderate AD. Although a phase I safety study in few individuals did not detect significant side-effects, a subsequent phase II trial was discontinued shortly after its initiation when roughly 5% of the treated participants developed what seemed to be an inflammatory reaction in the CNS (an aseptic meningoencephalitis). The occurrence of the meningocerebral inflammation was not correlated with either the presence or titres of antibodies against A among the trial participants. The mechanism of this self-limited inflammatory reaction is unknown, but the appearance of the inflammation before the detection of A antibodies in some of the recipients may suggest that a T-cell-mediated immune reaction to A was responsible. Such cellular reactions were not detected in mice and other mammals exposed to the vaccine during preclinical safety and efficacy testing, although a recent report suggests autoimmune encephalomyelitis can be induced in mice vaccinated with A peptide plus pertussis.

Efforts are underway to determine the basis for the adverse inflammatory reaction induced by A_{1-42} and to attempt to model it in animals. No abnormal effects have been documented in APP transgenic mouse models to which A antibodies have been administered, and such mice have shown robust clearing of brain A deposits and even improvements in behavioural deficits. This is in contrast to the EAE model in which administration of antibody to MOG worsens the progression of EAE. There is therefore an interest in conducting trials with a humanised monoclonal antibody to A as the next step in the clinical evaluation of the immunotherapeutic approach to AD. It may also be possible to immunise with portions of A to generate only antibodies that target N-terminal residues.

A as an autoantigen

We have found recently that APP transgenic mice, which produce robust quantities of A in the brain, have a form of immunological tolerance in which they show significantly lower T-cell responses when immunised with A than do wild-type mice. This deficit can be overcome in part by providing T-cell help to the animal. Thus, the presence of abundant A in the brain may not only cause local neuronal and glia damage but also hinder the generation of a therapeutic immune response, whether innate or induced.

Very recently, we have begun to extend such analyses to humans and, by using sensitive short-term cloning techniques, have found heightened *in vitro* reactivity of peripheral T-cells against A in some elderly individuals and people with AD. Early studies did not find lymphocyte proliferation in response to APP peptides in individuals with AD. The likelihood of seeing this T-cell hypereactivity in humans increased with age but was not observed in all individuals with AD or all aged normal individuals. Our results raise the possibility that endogenous T-cell reactivity in a host may relate to the progression of the cytopathological process of AD. In addition, such data suggest that it may useful to test individuals for their intrinsic T-cell reactivity to A before offering them any immunotherapeutic based on A.

BENEFICIAL VERSUS DELETERIOUS T-CELL RESPONSES

The issue of beneficial versus deleterious T-cell responses in vaccination models against CNS antigens is a concept that applies to approaches in both AD and MS. It has been shown that deleterious T-cell responses, presumably related to the induction of $T_{H}1$-type responses, can be induced in humans affected with either AD or MS.

This does not mean that vaccination approaches in CNS diseases cannot be successful, as has been shown by the use of glatiramer acetate in MS; however, strategies that induce nonpathogenic T-cell responses must be utilised, for example, modified autoantigens, tolerogenic routes such as mucosal administration and non-$T_{H}1$-inducing adjuvants should be used, and careful attention should be paid to dosing. In addition, the genetic background of the host and the immune repertoire may also determine whether a detrimental T-cell response will occur after vaccination.

For example, we have found that SJL mice strains immunised with MOG peptide in complete Freund's adjuvant are susceptible to EAE, whereas B10S mice treated similarly are resistant. This does not seem to relate to the generation of immune response against MOG, but to the type of immune response. In the SJL mouse there is infiltration of cells expressing -interferon in the brain and a predominantly $>_{H}1$ response, whereas in the B10S mice there is a $T_{H}2$ and $T_{H}3$ response that seems to prevent disease. Thus, the immune repertoire of the host before vaccination may determine the outcome of vaccination.

It seems that vaccination strategies both in AD and in MS will be dependent on skewing the immune response in such a way that it is not harmful to the host. In this regard, we have found in the APP mouse model of AD that nasal administration of A induces antibody responses in association with an 'anti-inflammatory' cellular immune response involving IL-4, IL-10 and TGF-. These 'anti-inflammatory' responses may themselves help the pathologic process by suppressing inflammation and microglial activation, which are believed to contribute to the CNS dysfunction in AD. Furthermore, cells secreting TGF- may themselves aid in the clearance of A. Such A-reactive T cells would act in the CNS only at sites where > is involved in the inflammatory process and thus would not be expected to interfere with normal physiology.

IMMUNOPATHOGENESIS OF SEPSIS

Sepsis describes a complex clinical syndrome that results from a harmful or damaging host response to infection. As a result of a concerted effort to understand the underlying pathogenetic mechanisms, there have been significant advances that have illuminated not just the process of sepsis, but also fundamental principles governing bacterial–host interactions. Unfortunately, attempts to translate these observations into improved clinical outcomes proved unsuccessful and led to considerable frustration. But in the past year, four major clinical trials that are based on somewhat different strategies have shown that it is possible to significantly reduce the mortality from sepsis and septic shock, and it is therefore timely to review these developments, both in basic science and its clinical applications.

Sepsis develops when the initial, appropriate host response to an infection becomes amplified, and then dysregulated. Clinically, the onset is often insidious: features may include fever, mental confusion, transient hypotension, diminished urine output or unexplained thrombocytopenia. If untreated, the patient may develop respiratory or renal failure, abnormalities of coagulation, and profound and unresponsive hypotension. A recent epidemiological study from North America found that the incidence was approximately 3.0 cases per 1,000 population, which translates into an annual burden of approximately 750,000 cases.

The overall mortality is approximately 30%, rising to 40% in the elderly and is 50% or greater in patients with the more severe syndrome, septic shock. It is worth emphasising that these figures represent mortality rates in patients admitted to hospital intensive care units and given antibiotics and the best available supportive care.

The commonest sites of infection are the lungs, abdominal cavity, the urinary tract and primary infections of the blood stream. A microbiological diagnosis is made in about half the cases; Gram-negative bacteria account for about 60% of cases, Gram-positive for the remainder.

MICROBIAL COMPONENTS THAT INITIATE INJURY

Determining the structural components of bacteria that are responsible for initiating the septic process has been important not only in understanding the underlying mechanisms, but also in identifying potential therapeutic targets. These bacterial motifs, which are recognised by the innate immune system, have been called pathogen-associated molecular patterns (PAMPs), although it might be more accurate to call them microorganism-associated molecular patterns as it is by no means clear how the host distinguishes between signals from pathogens rather than commensals.

In Gram-negative bacteria, lipopolysaccharide (LPS; known also as endotoxin) has a dominant role. The outer membrane of Gram-negative bacteria is constructed of a lipid bilayer, separated from the inner cytoplasmic membrane by peptidoglycan. The LPS molecule is embedded in the outer membrane and the lipid A portion of the molecule serves to anchor LPS in the bacterial cell wall.

Biophysical studies on the three-dimensional conformation adopted by different lipid A partial structures have revealed that, under physiological conditions, the most active forms assume the shape of a truncated cone, whereas inactive molecules prefer a lamellar structure and become progressively more cylindrical. These conformational changes seem to correlate with the ability to activate host cell membranes.

There is no endotoxin in Gram-positive bacteria, but their cell walls do contain peptidoglycan and lipoteichoic acid, and several investigators have identified structural components that account for their biological activity. Both peptidoglycan and lipoteichoic acid can bind to cell-surface receptors and are pro-inflammatory, although they are much less active, on a weight-for-weight basis, than LPS. Their role in the pathogenesis of clinical sepsis remains uncertain because there are no convincing clinical data to show that they are present in the circulation at concentrations comparable to those used in the experimental setting.

However, an important feature of Gram-positive cells is the production of potent exotoxins, some of which are implicated in septic shock. The best known examples are the toxic shock syndromes caused by toxic shock syndrome toxin-1 (TSST-1)-producing strains of *Staphylococcus aureus* and the pyrogenic exotoxins from *Streptococcus pyogenes*. Toxic shock syndromes are among the most acute and most severe forms of septic shock. They frequently occur without warning in otherwise healthy individuals and the mortality can be as high as 50%. These Gram-positive exotoxins are of great interest because they exhibit the properties of superantigens, that is, they are able to bind promiscuously to major histocompatibility complex class II and a restricted repertoire of T-lymphocyte receptor (TCR) V domains. In so doing they cause massive T-cell activation and release of pro-inflammatory lymphokines, suggesting a plausible role for these toxins as a cause of the profound shock that is seen in patients with toxic shock.

Detailed structural analyses have been done for many bacterial superantigens, and the crystal structures of several staphylococcal and streptococcal toxins have been elucidated. Interestingly, sequence variability in the amino-terminal domain dictates varying affinities for specific human leukocyte antigen (HLA) class II alleles; for instance, the streptococcal superantigen SPEA (for streptococcal pyrogenic exotoxin A) shows significantly greater affinity for HLA-DQ than HLA-DR. These differences may in part explain the remarkable selectivity of the toxic shock syndromes: although staphylococcal and streptococcal strains bearing superantigen genes are widespread and indeed frequently cause infections, toxic shock syndromes are relatively uncommon.

Although experimental and epidemiological studies provide some support for the view that these superantigenic toxins are the cause of the toxic shock syndromes, it is by no means clear that it is their superantigenicity *per se* that is responsible. For instance, despite many data that implicate the streptococcal toxin SPEA, this is in fact a relatively weak superantigen compared to the more recently described toxin streptococcal mitogenic exotoxin Z (SMEZ). Yet in experimental models in which HLA-DQ transgenic mice are challenged with strains of *S. pyogenes* in which *smez* is disrupted, there is no effect on survival despite a profound reduction in pro-inflammatory activity. These findings are important because there is considerable interest in devising therapeutic strategies that are targeted at Gram-positive infections and the toxic shock syndromes, and it is not clear whether these strategies should be aimed at the superantigenicity, or at other pro-inflammatory properties of the toxins.

There are also data that suggest that superantigenic toxins from Gram-positive bacteria induce hypersensitivity to LPS. The staphylococcal toxin TSST-1 enhances the susceptibility of rabbits to a lethal injection of LPS by a factor of approximately 50,000, and co-injection of LPS and TSST-1 induces tumour-necrosis factor- (TNF-) levels significantly higher than injection of similar doses of either toxin alone. Mice with severe combined immunodeficiency, lacking B and T lymphocytes, are resistant to this effect, but regain sensitivity when reconstituted with T-cells, and the mechanism seems to be dependent on enhanced production of interferon- (IFN-) from toxin-activated T cells. This interaction between superantigens and LPS might in part explain the devastating nature of the toxic shock syndromes. It could also have therapeutic implications, as it might be advantageous to target LPS even if the infection is apparently caused exclusively by Gram-positive bacteria.

Several other bacterial components have been shown to have pro-inflammatory activity and to be able to induce shock in experimental systems. These include cell-wall structures such as flagellin and curli, and unmethylated CpG sequences in naked bacterial DNA. Receptors for some of these elements have been identified among the family of Toll-like proteins that are now known to be crucial in the cellular recognition of microbial structures.

HOST RECOGNITION OF MICROBIAL COMPONENTS

The CD14–LBP complex The inability to identify an 'LPS receptor' was for many years a barrier to understanding how Gram-negative bacteria initiated the septic response, but in a series of elegant studies it was shown that activation of host cells was dependent on the presence of LPS-binding protein (LBP) and the opsonic receptor CD14. Although CD14 was originally identified as the essential co-receptor that mediated LPS activation of monocytes, subsequent work has shown that it also participates in the activation by Gram-positive cell-wall components such as peptidoglycan, mediates macrophage apoptosis, and is important in shuttling LPS between serum proteins that have the capacity to bind LPS, such as LBP and serum lipoproteins. Membrane bound CD14 (mCD14) is a glycosylphosphatidylinositol-linked molecule anchored in the cell surface, but it is also found in the circulation as soluble CD14 (sCD14). Many cells that are constitutively CD14 negative, such as dendritic cells, fibroblasts, smooth muscle cells and vascular endothelium, are still able to respond to LPS by interacting with sCD14. sCD14 is found in the serum of healthy individuals but levels rise in sepsis, and antibody to CD14 protects primates from lethal endotoxin shock.

Toll-like receptors

Although the discovery of CD14 represented a significant step forward in understanding host responses to LPS, the fact that mCD14 had no intracellular tail meant that it remained unclear how ligation of the LPS–LBP complex led to cellular activation. This uncertainty was resolved by the discovery of the family of Toll-like receptors (TLRs). Over a remarkably short period of time, studies of innate immunity in *Drosophila* revealed the existence of a proteolytic cascade that yielded ligands for cellular receptors that could distinguish bacterial from fungal infection. It was shown subsequently that there were striking similarities between this system and the interleukin (IL)-1 signalling system in mammals. This in turn led to the identification of human TLRs and the discovery that a TLR was the long-sought co-receptor for LPS.

A family of (currently) ten TLRs has been identified with a wide range of ligand specificity including bacterial, fungal and yeast proteins. Thus, TLR4 is the LPS receptor, TLR2 is predominantly responsible for recognising Gram-positive cell-wall structures, TLR5 is the receptor for flagellin and TLR9 recognises CpG elements in bacterial DNA. An additional cell-surface molecule, MD-2, has been identified that is required for activation of TLR4. MD-2 knockout mice do not respond to LPS and survive endotoxic shock. The role of MD-2 seems to be that of positioning TLR4 correctly on the cell surface, as in MD-$2^{-/-}$ embryonic fibroblasts TLR4 remained within the Golgi and failed to appear on the cell surface.

The notion of a 'monogamous' association between one particular TLR and its microbial ligand, as in the case of LPS and TLR4, is in reality an oversimplification.

For instance, TLR2 can be activated by cell-wall components of both yeast and mycobacteria. Further complexity is introduced into the system by the fact that TLRs seem to be able to combine to form a repertoire capable of distinguishing closely related ligands, and there is at least preliminary evidence that polymorphisms in Toll-family proteins might provide part of the explanation for the enormous variability in individual responses to what seem to be similar infective challenges.

Signalling pathways activated by TLRs have been dissected in great detail, and show a remarkable degree of homology with the Toll activation pathway in *Drosophila*. TLRs have an intracellular domain that is homologous with the IL-1 receptor and the IL-18 receptor. Adapter proteins facilitate binding to IL-1 receptor-associated kinase, which in turn induces TNF receptor-associated factor-6, leading to nuclear translocation of nuclear factor-B (NF-B) and ultimately to activation of cytokine gene promoters. Although this model is based on LPS signalling of TLR4, a similar — although not identical — process is involved in the activation of TLR2 by Gram-positive bacteria.

Other host signal molecules that respond to bacteria

A further layer of complexity has been provided by the discovery that there are several additional pathways by which cells recognise microbial components. Peptidoglycan-recognition proteins (PGRPs) were identified in moths and subsequently a family of PGRP genes was found in *Drosophila* and in humans.

Different PGRPs can distinguish between Gram-positive and Gram-negative bacteria. In *Drosophila*, they seem to act by regulating activation of Relish, a member of the NF-B family, although the precise mechanism by which they are sensed at the cell surface remains unknown.

The triggering receptor expressed on myeloid cells (TREM-1) and the myeloid DAP12-associating lectin (MDL-1) are two recently identified receptors involved in monocytic activation and inflammatory response. TREM-1 is upregulated in the presence of various microorganisms, although the ligand for TREM-1 is unknown. When mononuclear cells are exposed to a combination of LPS and an antibody to TREM-1, there is a synergistic effect and enhanced production of pro-inflammatory cytokines. But if a fusion protein of TREM-1 and the Fc portion of IgG is used to compete with cell-bound receptor, LPS-induced cytokine production is downregulated and mice can be protected from death up to 4 hours after a lethal injection of LPS. This is a therapeutic effect that will have obvious implications if it can be reproduced in clinical studies.

Finally, there is the recent description of the monocytic intracellular proteins NOD1 and NOD2 (for nucleotide-binding oligomerisation domain), which seem to have the ability to bind and to confer responsiveness to LPS, suggesting that this might be yet another way cells respond to the presence of bacteria. Genotypic variations in *NOD2* are associated with distinct clinical phenotypes of Crohn's disease, prompting speculation that other *NOD* genotypes might be associated with phenotypic variations in LPS responsiveness.

SIGNAL AMPLIFICATION

Following the initial host–microbial interaction there is widespread activation of the innate immune response, the purpose of which is to coordinate a defensive response involving both humoral and cellular components. Mononuclear cells play a key role, releasing the classic pro-inflammatory cytokines IL-1, IL-6 and TNF-, but in addition an array of other cytokines including IL-12, IL-15 and IL-18, and a host of other small molecules.

TNF- and IL-1 are the prototypic inflammatory cytokines that mediate many of the immunopathological features of LPS-induced shock. They are released during the first 30–90 minutes after exposure to LPS and in turn activate a second level of inflammatory cascades including cytokines, lipid mediators and reactive oxygen species, as well as upregulating cell adhesion molecules that result in the initiation of inflammatory cell migration into tissues.

The fact that anti-TNF or anti-IL-1 strategies failed to prevent death in septic patients is probably related more to the difficulty of designing clinical trials in these patients, rather than an intrinsic flaw in the scientific rationale. One practical problem is that patients often come to medical attention relatively late in the disease, and blocking these early cytokines may simply be too late. High mobility group B1 (HMGB1) has recently been identified as a cytokine-like product of macrophages that appears much later after LPS stimulation and may represent a more tractable target for intervention.

HMGB1 is a non-histone chromosomal protein that is abundantly distributed and exists in nuclear, cytoplasmic and membrane-bound forms. It participates in stabilising nucleosomes, facilitates gene transcription and modulates the activity of steroid hormone receptors. When mice were injected with LPS, HMGB1 serum concentrations rose after a delay of about 24 hours, long after the initial peak of IL-1 and TNF- had declined. Importantly, mice could be rescued from LPS-induced shock by administering an antibody to HMGB1, even when this was provided up to 2 hours after the lethal injection. Subsequently it was shown that patients with sepsis have elevated serum levels of HMGB1, and that higher levels are associated with an increased mortality, suggesting that clinical intervention by blocking or neutralising HMGB1 might be a viable option. Another macrophage-derived cytokine that has been identified as a potential therapeutic target in sepsis is macrophage migration inhibitory factor (MIF).

Mice with a targeted disruption of the MIF gene are resistant to LPS-induced shock and antibody to MIF is fully protective, even in the more demanding caecal ligation and puncture model that resembles clinical peritonitis. MIF also seems to mediate shock caused by Gram-positive bacteria, such as the toxic shock syndrome associated with *S. aureus*, suggesting that anti-MIF strategies might have broad application in septic patients. MIF has a curious relationship with glucocorticoids, which are normally thought of as being anti-inflammatory, as low doses of glucocorticoids paradoxically induce macrophage MIF. Once released, MIF then acts as a pro-inflammatory agent, over-riding the ability of glucocorticoids to prevent shock in animal models of sepsis. How this complex relationship manifests in a clinical setting is of particular interest in the light of the recent studies demonstrating a protective effect of low-dose steroids in patients with severe sepsis.

These pro-inflammatory cytokines are important because they in turn are responsible for orchestrating a complex network of secondary responses. A good example of this is provided by IL-18, a cytokine that induces production of interferon- (IFN-). In human mononuclear cells, IFN- upregulates surface expression of TLR4, MD-2 and MyD88, and counteracts the LPS-induced downregulation of TLR4. It has long been known that IFN- sensitises human mononuclear cells to the effects of LPS, and these new findings suggest strongly that this effect is probably mediated through upregulation (or at least, prevention of downregulation) of TLR4.

The coagulation cascade

Cytokines are also important in inducing a procoagulant effect in sepsis. Disorders of coagulation are common in sepsis, and 30–50% of patients have the more severe clinical form, disseminated intravascular coagulation. Coagulation pathways are initiated by LPS and other microbial components, inducing expression of tissue factor on mononuclear and endothelial cells. Tissue factor in turn activates a series of proteolytic cascades, which result in the conversion of prothrombin to thrombin, which in turn generates fibrin from fibrinogen. Simultaneously, normal regulatory fibrinolytic mechanisms (fibrin breakdown by plasmin) are impaired because of high plasma levels of plasminogen-activator inhibitor type-1 (PAI-1) that prevent the generation of plasmin from the precursor plasminogen. The net result is enhanced production and reduced removal of fibrin leading to the deposition of fibrin clots in small blood vessels, inadequate tissue perfusion and organ failure.

Pro-inflammatory cytokines, in particular IL-1 and IL-6, are powerful inducers of coagulation, and conversely, IL-10 regulates coagulation by inhibiting the expression of tissue factor on monocytes. An additional cause of the procoagulant state in sepsis is the downregulation of three naturally occurring anticoagulant proteins — antithrombin, protein C and tissue factor pathway inhibitor. These natural anticoagulants are of particular interest because in addition to their effect on thrombin generation, they also have anti-inflammatory properties, including effects on release of monocyte-derived TNF- by inhibiting activation of the transcription factors NF-B and activator protein (AP)-1.

Particular attention has focused on Protein C, which is converted to the activated form (aPC) when thrombin complexes with thrombomodulin, an endothelial transmembrane glycoprotein. Once aPC is formed it dissociates from an endothelial protein C receptor (EPCR) before binding protein S, resulting in inactivation of factors Va and VIIIa and thus blockade of the coagulation cascade. It has been shown recently that aPC uses EPCR as a co-receptor for cleavage of protease-activated receptor 1 (PAR1). Gene profiling showed that PAR1 signalling could account for the activation of aPC-induced protective genes, including the immunomodulatory monocyte chemoattractant protein-1 (MCP-1), suggesting a role for PAR-1 activation in protection from sepsis. In septic patients, aPC levels are reduced and expression of endothelial thrombomodulin and EPCR are impaired, providing some support for the notion that replacement of aPC might have therapeutic value.

The counter-inflammatory response

The profound pro-inflammatory response that occurs in sepsis is balanced by an array of counter-regulatory molecules that attempt to restore immunological equilibrium. In this sense, the counter-inflammatory response is seen as a 'modifier' — both appropriate and beneficial. Counter-inflammatory cytokines include antagonists such as the soluble TNF receptors and IL-1 receptor antagonist, decoy receptors such as IL-1 receptor type II, inactivators of the complement cascade and the anti-inflammatory cytokines, of which the prototype is IL-10.

In concert with this, the host response to injury includes profound changes in metabolic activity (increased cortisol production and release of catecholamines), induction of acute-phase proteins, and endothelial activation with upregulation of adhesion molecules and release of prostanoids and platelet-activating factor (PAF).

Another facet of downregulation of immunity that occurs in sepsis is the development of lymphocyte apoptosis. Extensive lymphocyte apoptosis is seen in animal models of sepsis and is also present in septic patients, although interestingly, much less so in critically ill non-septic controls. Septic patients are usually lymphopenic, and subset analysis of autopsy tissue samples has shown that there is selective depletion of B and $CD4^+$ lymphocytes. This process and its functional consequences are viewed as part of a more general state of immunosuppression, characterised by T-cell hyporesponsiveness and anergy, which occurs to some extent in most septic patients, and which is seen as a counter-balancing response (and sometimes, over-response) to the initial pro-inflammatory state.

It is because of this over-response that some investigators view the counter-inflammatory response as the cause of an inadequate host defence against infection and hence a potential 'mediator' of sepsis and progressive organ failure.

Several have pursued the notion that reversal of this immunosuppressive state might be of therapeutic value. For instance, mice transfected with the human gene *bcl-2* that overexpress the anti-apoptotic protein Bcl-2 are protected from death after caecal ligation and puncture, and patients that received IFN- in a small non-randomised clinical study showed upregulation of HLA-DR on their monocytes and a better-than-anticipated survival.

ROLE OF GENETIC SUSCEPTIBILITY IN THE PATHOGENESIS OF SEPSIS

Among this vast array of host molecules that orchestrate the response to sepsis there are many examples of genetic variability that influence physiological activity. For example, there has been great interest in exploring the possibility that a polymorphism in the TNF promoter that results in significantly higher TNF levels might be associated with a worse outcome from sepsis. Several of these associations have been studied and at least in some cases the evidence seems convincing. Of particular interest was the recent report that mutations in TLR4 are associated with an increased susceptibility to Gram-negative sepsis.

Mechanisms of organ failure

The ultimate cause of death in patients with sepsis is multiple organ failure. Typically, patients will first develop a single organ failure — for instance, respiratory failure requiring mechanical ventilation — and then if the disease remains unchecked, will progressively develop failure of other organ systems. There is a close relationship between the severity of organ dysfunction on admission to an intensive care unit and the probability of survival, and between the numbers of organs failing and the risk of death. If four or five organs fail the mortality is greater than 90%, irrespective of treatment.

The pathogenesis of organ dysfunction is multifactorial and incompletely understood. Tissue hypoperfusion and hypoxia are dominant factors. The mechanisms involve widespread fibrin deposition causing microvascular occlusion, the development of tissue exudates further compromising adequate oxygenation, and disorders of microvascular homeostasis resulting from the elaboration of vasoactive substances such as PAF, histamine and prostanoids. Cellular infiltrates, particularly neutrophils, damage tissue directly by releasing lysosomal enzymes and superoxide-derived free radicals. TNF- and other cytokines increase the expression of the inducible nitric oxide synthase and increased production of nitric oxide causes further vascular instability and may also contribute to the direct myocardial depression that occurs in sepsis.

The tissue hypoxia that develops in sepsis is reflected in the oxygen debt — that is, the difference between oxygen delivery and oxygen requirements. The extent of the oxygen debt is related to the outcome from sepsis, and strategies designed to optimise oxygen delivery to the tissues can improve survival. In addition to hypoxia, cells may be dysoxic — that is, unable to properly utilise available oxygen. Recent data suggest that this may be another consequence of excess nitric oxide production, because skeletal muscle biopsies from septic patients show evidence of impaired mitochondrial respiration, which is inhibited by nitric oxide.

Therapeutic approaches

Despite the extraordinary developments in understanding the immunopathology of sepsis, therapeutic advances have been painfully slow. However, in the past 12 months several clinical trials have finally shown that it is possible to reduce mortality in patients with sepsis. Interestingly, each has been based on a different aspect of the pathology.

The critical importance of tissue oxygenation was addressed by a study in which patients in the earliest stages of sepsis were treated by aggressive management with fluids, blood transfusion and inotropic agents to optimise haemodynamic function. An alternative approach was taken by van den Berghe and co-workers, who studied the effect of rigorous control of blood glucose levels. Hyperglycaemia and insulin resistance are common in critically ill patients, even if they have not previously had diabetes, and these authors showed that intensive insulin therapy could substantially reduce mortality. The mechanism of this striking effect is not absolutely clear, although it is of interest that the greatest reduction in mortality involved deaths due to multiple-organ failure in patients with a proven septic focus, perhaps suggesting that it was related to the better control of the initial infective process.

The third study to show significant benefit was a trial of low-dose corticosteroids. Earlier trials using very high doses of steroids, based on the premise that sepsis represented an uncontrolled inflammatory response, had failed to show any survival benefit. But Annane and colleagues had noted that patients in the advanced stages of septic shock had relative adrenal insufficiency and reasoned that low-dose replacement steroids might be beneficial. In a phase III trial in highly selected patients, they found that low doses of hydrocortisone and fludrocortisone did indeed reduce the mortality substantially. But doubts remain about the precise mechanism of this effect and it is possible that the benefit derives, at least in part, from the immunosuppressive effects of the hydrocortisone.

Disordered coagulation is important in sepsis, and the fourth study examined the effects on the course of sepsis of replacing aPC. A randomised controlled trial of aPC therapy resulted in a significant survival benefit in treated patients, although interestingly the effect was equally marked in patients whose protein C levels were not depressed. This suggests that the therapeutic effect was attributable in part to the immunosuppressive effects of protein C, and not just to its anticoagulant properties.

Improved understanding of the immunopathology of sepsis has facilitated many other approaches, and several additional strategies are at various stages of development. These include therapies aimed at bacterial targets, for example novel anti-endotoxin molecules such as bactericidal/permeability-increasing protein, or modified lipoproteins, both of which absorb and neutralise LPS, as well as very recent reports that oxidised phospholipids can interfere with binding of LPS to LBP, and strategies aimed at Gram-positive toxins, including competitive antagonists of superantigen-binding sites. There are also investigations aimed at host molecules, such as PAF-receptor antagonists, and a variety of targets in the coagulation cascade. For the clinical investigator, the challenge will be the design of studies with sufficient power to determine the potential value of these new therapies. As fatality rates drop with the use of low-dose steroids, aPC and the other measures, it will become increasingly difficult to carry out studies on a heterogeneous population of patients with sepsis. The way forward is likely to lie in identifying clinically relevant endpoints other than death (for instance, reduced incidence of organ failure), and/or identifying more homogeneous subgroups of patients in whom to study specifically targeted therapeutic interventions.

5

Immunology and Immunopathology

Over the past decade, there have been numerous advances in our current understanding of the immune system and how it functions to protect the body from infection. Given the complex nature of this subject, it is beyond the scope of this article to provide an in-depth review of all aspects of immunology. Rather, the purpose of this article is to provide medical students, medical residents, primary-care practitioners and other health care professionals with a basic introduction to the main components and function of the immune system and its role in both health and disease. This article will also serve as a backgrounder to the immunopathological disorders discussed in the remainder of this supplement. The topics covered in this introductory article include: innate and acquired immunity, passive and active immunisation and immunopathologies, such as hypersensitivity reactions, autoimmunity and immunodeficiency.

THE IMMUNE SYSTEM: INNATE AND ADAPTIVE IMMUNITY

The immune system refers to a collection of cells and proteins that function to protect the skin, respiratory passages, intestinal tract and other areas from foreign antigens, such as microbes (organisms such as bacteria, fungi, and parasites), viruses, cancer cells, and toxins. The immune system can be simplistically viewed as having two "lines of defence": innate immunity and adaptive immunity. Innate immunity represents the first line of defence to an intruding pathogen. It is an antigen-independent (non-specific) defence mechanism that is used by the host immediately or within hours of encountering an antigen.

The innate immune response has no immunologic memory and, therefore, it is unable to recognise or "memorize" the same pathogen should the body be exposed to it in the future. Adaptive immunity, on the other hand, is antigen-dependent and antigen-specific and, therefore, involves a lag time between exposure to the antigen and maximal response. The hallmark of adaptive immunity is the capacity for memory which enables the host to mount a more rapid and efficient immune response upon subsequent exposure to the antigen. Innate and adaptive immunity are not mutually exclusive mechanisms of host defence, but rather are complementary, with defects in either system resulting in host vulnerability.

INNATE IMMUNITY

The primary function of innate immunity is the recruitment of immune cells to sites of infection and inflammation through the production of cytokines (small proteins involved in cell-cell communication). Cytokine production leads to the release of antibodies and other proteins and glycoproteins which activate the complement system, a biochemical cascade that functions to identify and opsonize (coat) foreign antigens, rendering them susceptible to phagocytosis (process by which cells engulf microbes and remove cell debris). The innate immune response also promotes clearance of dead cells or antibody complexes and removes foreign substances present in organs, tissues, blood and lymph. It can also activate the adaptive immune response through a process known as antigen presentation.

Numerous cells are involved in the innate immune response such as phagocytes (macrophages and neutrophils), dendritic cells, mast cells, basophils, eosinophils, natural killer (NK) cells and lymphocytes (T cells). Phagocytes are sub-divided into two main cell types: neutrophils and macrophages. Both of these cells share a similar function: to engulf (phagocytose) microbes. In addition to their phagocytic properties, neutrophils contain granules that, when released, assist in the elimination of pathogenic microbes. Unlike neutrophils (which are short-lived cells), macrophages are long-lived cells that not only play a role in phagocytosis, but are also involved in antigen presentation to T cells. Macrophages are named according to the tissue in which they reside. For example, macrophages present in the liver are called Kupffer cells while those present in the connective tissue are termed histiocytes.

Dendritic cells also phagocytose and function as antigen-presenting cells (APCs) and act as important messengers between innate and adaptive immunity. Mast cells and basophils share many salient features with each other and both are instrumental in the initiation of acute inflammatory responses, such as those seen in allergy and asthma.

Unlike mast cells, which generally reside in the connective tissue surrounding blood vessels, basophils reside in the circulation. Eosinophils are granulocytes that possess phagocytic properties and play an important role in the destruction of parasites that are too large to be phagocytosed. Along with mast cells and basophils, they also control mechanisms associated with allergy and asthma. NK cells (also known as large granular lymphocytes [LGLs]) play a major role in the rejection of tumours and the destruction of cells infected by viruses.

Destruction of infected cells is achieved through the release of perforins and granzymes from NK-cell granules which induce apoptosis (programmed cell death). The main characteristics and functions of the cells involved in the innate immune response.

Innate immunity can be viewed as comprising four types of defensive barriers: anatomic (skin and mucous membrane), physiologic (temperature, low pH and chemical mediators), endocytic and phagocytic, and inflammatory.

Adaptive immunity

Adaptive immunity develops when innate immunity is ineffective in eliminating infectious agents and the infection is established. The primary functions of the adaptive immune response are the recognition of specific "non-self" antigens in the presence of "self" antigens; the generation of pathogen-specific immunologic effector pathways that eliminate specific pathogens or pathogen-infected cells; and the development of an immunologic memory that can quickly eliminate a specific pathogen should subsequent infections occur. The cells of the adaptive immune system include: T cells, which are activated through the action of antigen presenting cells (APCs), and B cells.

T cells and APCs

T cells derive from hematopoietic stem cells in bone marrow and, following migration, mature in the thymus. These cells express a unique antigen-binding receptor on their membrane, known as the T-cell receptor (TCR), and as previously mentioned, require the action of APCs (usually dendritic cells, but also macrophages, B cells, fibroblasts and epithelial cells) to recognise a specific antigen.

The surfaces of APCs express cell-surface proteins known as the major histocompatibility complex (MHC). MHC are classified as either class I (also termed human leukocyte antigen [HLA] A, B and C) which are found on all nucleated cells, or class II (also termed HLA, DP, DQ and DR) which are found on only certain cells of the immune system, including macrophages, dendritic cells and B cells. Class I MHC molecules present endogenous (intracellular) peptides while class II molecules present exogenous (extracellular) peptides. The MHC protein displays fragments of antigens (peptides) when a cell is infected with a pathogen or has phagocytosed foreign proteins.

T cells are activated when they encounter an APC that has digested an antigen and is displaying antigen fragments bound to its MHC molecules. The MHC-antigen complex activates the TCR and the T cell secretes cytokines which further control the immune response. This antigen presentation process stimulates T cells to differentiate into either cytotoxic T cells (CD8+ cells) or T-helper (Th) cells (CD4+ cells). Cytotoxic T cells are primarily involved in the destruction of cells infected by foreign agents. They are activated by the interaction of their TCR with peptide-bound MHC class I molecules. Clonal expansion of cytotoxic T cells produce effector cells which release perforin and granzyme (proteins that causes lysis of target cells) and granulysin (a substance that induces apoptosis of target cells). Upon resolution of the infection, most effector cells die and are cleared by phagocytes. However, a few of these cells are retained as memory cells that can quickly differentiate into effector cells upon subsequent encounters with the same antigen.

T helper (Th) cells play an important role in establishing and maximising the immune response. These cells have no cytotoxic or phagocytic activity, and cannot kill infected cells or clear pathogens. However, they "mediate" the immune response by directing other cells to perform these tasks. Th cells are activated through TCR recognition of antigen bound to class II MHC molecules. Once activated, Th cells release cytokines that influence the activity of many cell types, including the APCs that activate them.

Two types of Th cell responses can be induced by an APC: Th1 or Th2. The Th1 response is characterised by the production of interferon-gamma (IFN-□) which activates the bactericidal activities of macrophages, and other cytokines that induce B cells to make opsonising (coating) and neutralising antibodies. The Th2 response is characterised by the release of cytokines (interleukin-4, 5 and 13) which are involved in the activation and/or recruitment of immunoglobulin E (IgE) antibody-producing B cells, mast cells and eosinophils. As mentioned earlier, mast cells and eosinophils are instrumental in the initiation of acute inflammatory responses, such as those seen in allergy and asthma. IgE antibodies are also associated with allergic reactions. Therefore, an imbalance of Th2 cytokine production is associated with the development of atopic (allergic) conditions. Like cytotoxic T cells, most Th cells will die upon resolution of infection, with a few remaining as Th memory cells.

A third type of T cell, known as the regulatory T cell (T reg), also plays a role in the immune response. T reg cells limit and suppress the immune system and, thereby, may function to control aberrant immune responses to self-antigens and the development of autoimmune disease.

B cells

B cells arise from hematopoietic stem cells in the bone marrow and, following maturation, leave the marrow expressing a unique antigen-binding receptor on their membrane. Unlike T cells, B cells can recognise free antigen directly, without the need for APCs. The principal function of B cells is the production of antibodies against foreign antigens.

When activated by foreign antigens, B cells undergo proliferation and differentiate into antibody-secreting plasma cells or memory B cells. Memory B cells are "long-lived" survivors of past infection and continue to express antigen-binding receptors. These cells can be called upon to respond quickly and eliminate an antigen upon re-exposure. Plasma cells, on the other hand, do not express antigen-binding receptors. These are short-lived cells that undergo apoptosis when the inciting agent that induced the immune response is eliminated.

Given their function in antibody production, B cells play a major role in the humoral or antibody-mediated immune response (as opposed to the cell-mediated immune response, which is governed primarily by T cells).

AUTOIMMUNITY AND IMMUNE COMPLEX DISEASES

Immune disorders can be classified in several ways: by the source of the inducing, or reacting, antigen (heterologous, homologous, autologous); by the model hypersensitivity reactions types I-IV (IgE-, antibody-, immune complex-, or cell-mediated); and by disease expression characterised by clinical and immunological features, such as allergy/hypersensitivity, autoimmune disease, immune deficiency, and immunoproliferative disorder.

A central function of the immune system is to distinguish foreign antigens, such as infectious agents, from self components of body tissues. The immune system normally acquires self tolerance (unresponsiveness to self) by clonal deletion of autoreactive T cells in the thymus in the perinatal period and by functional suppression of autoreactive T and B cells at later stages of development. Nevertheless, sometimes there is a failure in the maintenance of self tolerance, a failure to discriminate between self and non-self antigens, and an autoimmune response, characterised by the activation and clonal expansion of autoreactive lymphocytes and the production of autoantibodies, is produced against autologous antigens of normal body tissues.

While there are many laboratory animal models of autoimmunity, there is currently no general unifying theory to explain how an autoimmune process gets started in the usual clinical setting. Suffice it to say that autoimmunity and autoimmune diseases are multifactorial in origin. Contributory factors include: genetic predisposition (as shown by HLA associations, family, and twin studies), host factors (such as weakness of immunoregulatory controls, defects in suppressor T cells, or polyclonal stimulation of B cells resistant to controls), environmental factors (such as certain microbial infections), and antigen-driven mechanisms (sequestered antigen, cross-reacting exogenous antigen, or molecular mimicry) which may bypass self tolerance in a nominally normal immune system. Current investigations of several autoimmune diseases are focused on the concept of molecular mimicry by which a viral, bacterial, or other foreign antigen bearing antigenic epitopes with aminoacid sequence homology to epitopes on a normal host protein elicits, in responsive individuals, the production of antibodies and activated T cells that react with the host protein as well as with the foreign protein.

Analogous to the manner in which hypersensitivity reactions to exogenous antigens initiate tissue injury and inflammation, so also autoantibody-, immune complex (IC)- or cell-mediated reactions to autologous antigens can lead to tissue injury and inflammation, resulting in an autoimmune disease. However, not all autoantibodies are pathogenic: some are, some are not, and some autoantibodies are the result, rather than the cause, of tissue injury. This is shown, for example, by the transient appearance of myocardial autoantibodies in some patients after myocardial infarction or after invasive heart surgery.

Autoimmune diseases, mainly those in which autoimmunity contributes to, or has an association with, the pathogenesis of the disease, can be classified into two broad, but overlapping, groups: organ-specific and nonorgan-specific (or systemic) autoimmune diseases. In the first mentioned, local injury, inflammation, or dysfunction are produced by autoantibody- or cell-mediated reactions against a specific target antigen located in a specialised cell, tissue, or organ. Clinical examples include: autoimmune hemolytic anemia (erythrocyte autoantibodies); Hashimoto's thyroiditis (thyroid autoantibodies and autoreactive T cells); myasthenia gravis (acetylcholine receptor autoantibodies); Grave's disease characterised by diffuse goiter and hyperthyroidism (thyrotropin receptor autoantibodies);

Goodpasture's syndrome comprising anti-GBM nephritis and pulmonary intraalveolar hemorrhage (anti-GBM autoantibodies); and type I (insulin - dependent) diabetes (pancreatic beta-cell autoreactive T cells and autoantibodies).

In a systemic autoimmune disease, by contrast, tissue injury and inflammation occur in multiple sites in organs without relation to their antigenic makeup and are usually initiated by the vascular leakage and tissue deposition of circulating autologous immune complexes. These ICs are formed by autoantibody responses to ubiquitous soluble cellular antigens of nuclear, or less commonly cytoplasmic, origin. Systemic lupus erythematosus (SLE), which is characterised by the production of multiple autoantibodies, particularly antinuclear and anti-DNA antibodies, is the classical example of an IC-mediated systemic autoimmune disease.

IMMUNE COMPLEX DISEASES

SERUM SICKNESS

Historically at the beginning of the 20th century, children with life-threatening bacterial infections, such as diphtheria, were sometimes treated with a large injection of horse serum antitoxin. After a latent period of 8-12 days, some of the children developed a self-limited syndrome called serum sickness and consisting of fever, urticaria, muscle and joint pains, lymphadenopathy, and proteinuria.

While the clinical use of heterologous serum is rare today and essentially limited to horse antirabies serum, tetanus antitoxin, and antisnake venum, the contemporary relevance of serum sickness is that the elucidation of its pathogenesis as a hypersensitivity reaction to foreign serum protein and an IC-mediated disease has advanced understanding of human glomerular and connective-tissue diseases.

Rabbit model of serum sickness

A laboratory model of serum sickness can be produced in rabbits by a one-shot i.v. injection of a foreign serum protein, such as BSA (bovine serum albumin). Blood serum and tissue specimens are taken at various intervals to determine the time-course of immunopathological events.

Time-course of immunopathological events in rabbit model of serum sickness induced by one-shot i.v. injection of bovine serum albumin (BSA) at time zero. After an interval of a week or so, the immune phase of BSA antigen elimination begins as newly produced anti-BSA antibody forms circulating BSA-anti-BSA complexes in an antigen-excess environment. The resulting expression of immune complex-initiated tissue injury and inflammation, as shown for glomerulonephritis, first appears as the immune phase begins, reaches greatest incidence (and severity) about the time the immune phase is completed, and decreases following the elimination of all complexes and the appearance of free antibody.

The injected BSA is removed from the circulation in three stages : equilibration; catabolism; and, after a latent period of a week or so, the immune phase of antigen elimination during which anti-BSA antibody is produced in an antigen excess environment, forms soluble BSA-antiBSA complexes, and the circulating ICs are rapidly removed by the mononuclear phagocytic system. Thereafter, free antibody appears in the circulation.

Manifestations of serum sickness, expressed as tissue injury and acute inflammation in multiple sites, develop during the immune phase of antigen elimination: glomerulonephritis (GN), vasculitis, synovitis, and endocarditis.

The development and progression of the lesions coincide with the presence of soluble ICs in the circulation, a decrease in serum C' activity, and the tissue deposition of ICs and C' in the involved sites. The lesions begin to regress and heal after all complexes are eliminated and free antibody appears in the circulation.

The inflammatory lesions in experimental serum sickness are initiated by (type III) IC-mediated mechanisms. The pathogenic complexes are those produced in moderate antigen excess, tend to form antibody doublets (Ag3Ab2), have molecular masses < 1,000,000, and when deposited at sites of vascular permeability activate the classical C'-, neutrophil-, and macrophage-mediated mechanisms of tissue injury and inflammation. Vascular permeability is increased by platelet-derived amines and by C3a and C5a. Neutrophils are attracted by C5a. The GBM and other small vessel walls are injured by collagenases, proteases, other enzymes, and inflammatory mediators released from recruited neutrophils and macrophages and by the C5b- 9 membrane attack complex (MAC), resulting in vasculitis, GN, synovitis, serositis, and endocarditis. By electron microscopy, corresponding electron dense deposits are seen along the GBM mainly as subendothelial deposits.

The rabbit model of acute serum sickness can be transformed into chronic serum sickness if daily injections of BSA are made in order to maintain ICs in the circulation for an extended period. The expression of systemic IC disease then becomes chronic and progressive.

Other models of IC-mediated disease

Natural animal models of systemic IC disease include those induced by replicating foreign antigens, as with persistent viral infections, or by ever present autoantigens, as with autoimmune-disease prone strains of laboratory mice.

The Arthus reaction, the classical model of local IC disease, is produced experimentally by the intracutaneous injection of antigen in previously sensitized animals which already have precipitating antibodies to the antigen. As antigen diffuses into an antibody excess environment, large ICs are precipitated within the blood vessel walls, initiating C'- and neutrophil-mediated mechanisms of local injury and inflammation usually expressed as acute necrotizing vasculitis of the skin.

It can also be noted that the immune mechanisms of glomerular injury include not only type III hypersensitivity but also antibody-mediated (type II) hypersensitivity resulting from the in-situ formation of ICs of specific antibody with tissue localised antigen, in the absence of circulating ICs. The target antigen may be an intrinsic glomerular antigen, such as the GBM, as epitomised by anti-GBM nephritis in which a tissue-specific anti-GBM autoantibody binds to the GBM and produces a characteristic linear immunofluorescence pattern of Ig and C' deposits along the entire length of the GBM, contrasting with the granular pattern typical of deposits of circulating ICs.

Or, the target antigen for (type II) antibody-mediated injury may be a nonglomerular antigen of exogenous origin (such as an infectious agent) or endogenous derivation (such as DNA in individuals with SLE) which becomes bound or "planted" in the glomeruli because of affinity for components of the GBM. With a planted antigen, the IC deposits along the GBM produce an interrupted granular immunofluorescence pattern.

HUMAN IMMUNE COMPLEX DISEASES

Immune complex diseases are those in which the tissue deposition of circulating ICs initiates tissue injury and inflammation in multiple sites. Serum sickness is a prototype. Many human diseases are associated with circulating ICs, and some of them show granular deposits of Igs and C' in the inflammatory lesions of GN, vasculitis, and so on. The evoking antigen may be exogenous or autologous or is unknown in the majority of cases. The main IC-related human diseases are: the diffuse connective tissue diseases exemplified by SLE, a prototype systemic autoimmune disease; GN and vasculitis resulting from immune responses induced by microbial, autologous, or unknown antigens.

ANTIBODY-MEDIATED VS. CELL-MEDIATED IMMUNITY

Antibody-mediated immunity is the branch of the acquired immune system that is mediated by B-cell antibody production. The antibody-production pathway begins when the B cell's antigen-binding receptor recognises and binds to antigen in its native form. This, in turn, attracts the assistance of Th cells which secrete cytokines that help the B cell multiply and mature into antibody-secreting plasma cells.

The secreted antibodies bind to antigens on the surface of pathogens, flagging them for destruction through pathogen and toxin neutralisation, classical complement activation, opsonin promotion of phagocytosis and pathogen elimination. Upon elimination of the pathogen, the antigen-antibody complexes are cleared by the complement cascade.

Five types of antibodies are produced by B cells: immunoglobulin A (IgA), IgD, IgE, IgG and IgM. Each of these antibodies has differing biological functions and recognise and neutralise specific pathogens.

Antibodies play an important role in containing virus proliferation during the acute phase of infection. However, they are not generally capable of eliminating a virus once infection has occurred. Once an infection is established, cell-mediated immune mechanisms are most important in host defence.

Cell-mediated immunity does not involve antibodies, but rather protects an organism through :

- the activation of antigen-specific cytotoxic T cells that induce apoptosis of cells displaying epitopes (localised region on the surface of an antigen that is capable of eliciting an immune response) of foreign antigen on their surface, such as virus-infected cells, cells with intracellular bacteria, and cancer cells displaying tumour antigens;
- the activation of macrophages and NK cells, enabling them to destroy intracellular pathogens; and
- the stimulation of cytokine production that further mediates the immune response.

Cell-mediated immunity is directed primarily at microbes that survive in phagocytes as well as those that infect non-phagocytic cells. This type of immunity is most effective in eliminating virus-infected cells, but can also participate in defending against fungi, protozoa, cancers, and intracellular bacteria. Cell-mediated immunity also plays a major role in transplant rejection.

PASSIVE VS. ACTIVE IMMUNISATION

Acquired immunity is attained through either passive or active immunisation. Passive immunisation refers to the transfer of *active* humoral immunity, in the form of "ready-made" antibodies, from one individual to another.

It can occur naturally by transplacental transfer of maternal antibodies to the developing fetus, or it can be induced artificially by injecting a recipient with exogenous antibodies targeted to a specific pathogen or toxin. The latter is used when there is a high risk of infection and insufficient time for the body to develop its own immune response, or to reduce the symptoms of chronic or immunosuppressive diseases.

Active immunisation refers to the production of antibodies against a specific agent *after* exposure to the antigen. It can be acquired through either natural infection with a microbe or through administration of a vaccine that can consist of attenuated (weakened) pathogens or inactivated organisms,

IMMUNOPATHOLOGY

As mentioned earlier, defects or malfunctions in either the innate or adaptive immune response can provoke illness or disease. Such disorders are generally caused by an overactive immune response (known as hypersensitivity reactions), an inappropriate reaction to self (known as autoimmunity) or ineffective immune responses (known as immunodeficiency).

Hypersensitivity reactions

Hypersensitivity reactions refer to undesirable responses produced by the normal immune system. There are four types of hypersensitivity reactions :

- Type I: immediate hypersensitivity
- Type II: cytotoxic or antibody-dependent hypersensitivity
- Type III: immune complex disease
- Type IV: delayed-type hypersensitivity

Type I hypersensitivity is the most common type of hypersensitivity reaction. It is an allergic reaction provoked by re-exposure to a specific type of antigen, referred to as an allergen. Unlike the normal immune response, the type I hypersensitivity response is characterised by the secretion of IgE by plasma cells. IgE antibodies bind to receptors on the surface of tissue mast cells and blood basophils, causing them to be "sensitised". Later exposure to the same allergen, cross-links the bound IgE on sensitised cells resulting in degranulation and the secretion of active mediators such as histamine, leukotriene, and prostaglandin that cause vasodilation and smooth-muscle contraction of the surrounding tissue. Common environmental allergens inducing IgE-mediated allergies include cat-, dog- and horse epithelium, pollen, house dust mites and molds. Food allergens are also a common cause of type I hypersensitivity reactions, however, these types of reactions are more frequently seen in children than adults. Treatment of type I reactions generally involves trigger avoidance, and in the case of inhaled allergens, pharmacological intervention with bronchodilators, antihistamines and anti-inflammatory agents. More severe cases may be treated with immunotherapy.

Type II hypersensitivity reactions are rare and take anywhere from 2 to 24 hours to develop. These types of reactions occur when IgG and IgM antibodies bind to the patient's own cell-surface molecules, forming complexes that activate the complement system. This, in turn, leads to opsonisation, red blood cell agglutination (process of agglutinating or "clumping together"), cell lysis and death. Some examples of type II hypersensitivity reactions include: erythroblastosis fetalis, Goodpasture's syndrome, and autoimmune anemias.

Type III hypersensitivity reactions occur when IgG and IgM antibodies bind to soluble proteins (rather than cell surface molecules as in type II hypersensitivity reactions) forming immune complexes that can deposit in tissues, leading to complement activation, inflammation, neutrophil influx and mast cell degranulation.

This type of reaction can take hours, days, or even weeks to develop and treatment generally involves anti-inflammatory agents and corticosteroids. Examples of type III hypersensitivity reactions include systemic lupus erythematosus (SLE), serum sickness and reactive arthritis.

Unlike the other types of hypersensitivity reactions, type IV reactions are cell-mediated and antibody-independent. They are the second most common type of hypersensitivity reaction and usually take 2 or more days to develop. These types of reactions are caused by the overstimulation of T cells and monocytes/macrophages which leads to the release of cytokines that cause inflammation, cell death and tissue damage. In general, these reactions are easily resolvable through trigger avoidance and the use of topical corticosteroids.

Autoimmunity

Autoimmunity involves the loss of normal immune homeostasis such that the organism produces an abnormal response to its own tissue. The hallmark of autoimmunity is the presence of self-reactive T cells, auto-antibodies, and inflammation. Prominent examples of autoimmune diseases include: Celiac disease, type 1 diabetes mellitus, Addison's disease and Graves' disease.

Immunodeficiency

Immunodeficiency refers to a state in which the immune system's ability to fight infectious disease is compromised or entirely absent. Immunodeficiency disorders may result from a primary congenital defect (primary immunodeficiency) or may be acquired from a secondary cause (secondary immunodeficiency), such as viral or bacterial infections, malnutrition or treatment with drugs that induce immunosuppression. Certain diseases can also directly or indirectly impair the immune system such as leukemia and multiple myeloma. Immunodeficiency is also the hallmark of acquired immunodeficiency syndrome (AIDS), caused by the human immunodeficiency virus (HIV). HIV directly infects Th cells and also impairs other immune system responses indirectly.

HUMAN DISEASES ASSOCIATED WITH IMMUNE COMPLEXES

DIFFUSE CONNECTIVE TISSUE DISEASES (CTDS)

Autoimmune Diseases:

- Systemic Lupus Erythematosus (SLE)
- Rheumatoid Arthritis (RA)
- Sjogren's Syndrome (SS)
- Progressive Systemic Sclerosis (PSS)
- Mixed Connective Tissue Disease (MCTD)

Other:

- Polyarteritis Nodosa (PAN)

GLOMERULONEPHRITIS (GN)

- Exogenous antigens (bacterial, viral, parasitic)
- Autologous antigens (nuclear, IgG, thyroglobulin, tumor)

INFECTIOUS DISEASES

- Bacterial: infective endocarditis, disseminated streptococcal, staphylococcal, meningococcal, gonococcal infections, Lyme disease (borreliosis), syphilis, leprosy
- Viral: hepatitis B, cytomegalovirus infection, infectious mononucleosis
- Parasitic: malaria, toxoplasmosis, trypanosomiasis

Diffuse Connective Tissue Diseases

This diverse group of diseases of unknown etiologies and overlapping clinical and laboratory features was originally brought together under the descriptive term "diffuse collagen diseases" or "collagen vascular diseases" and is now generally called the diffuse connective tissue diseases (CTDs) or sometimes the autoimmune CTDs.

The CTDs include:

- rheumatoid arthritis (RA), a chronic systemic inflammatory disease characterised by polyarthritis (simultaneous inflammation of several joints) which may lead to joint destruction and deformity;
- systemic lupus erythematosus (SLE), a multisystem inflammatory disease involving many organs (skin, joints, kidneys);
- scleroderma/ progressive systemic sclerosis (PSS) characterised by dermal fibrosis and skin thickening (scleroderma) and progressive fibrosis (systemic sclerosis) of internal organs (GI tract, kidneys, heart, lungs) ;

- Sjogren's syndrome (SS) characterised by a syndrome of dry mouth/ dry eyes (due to lymphocytic infiltration of salivary and lacrymal glands), polyarthritis, and either a primary disease or secondary to RA or SLE;
- polymyositis/ dermatomyositis, primary inflammatory disorders of striated muscle, also with involvement of skin (dermatomyositis) and internal organs (heart, lungs);
- mixed connective tissue disease (MCTD) with overlapping clinical features of SLE, scleroderma, and polymyositis; and
- polyarteritis nodosa (PAN), a noninfectious necrotizing vasculitis of small and medium sized arteries of almost any organ (kidneys, heart, intestine).

Rheumatic fever was once included in this group but is now established as a poststreptococcal hypersensitivity disease.

Several of the CTDs are associated with antinuclear antibodies, anti-IgG antibodies called rheumatoid factors, and other autoantibodies. IC-mediated mechanisms are involved in the pathogenesis of SLE and, although less compellingly, RA and PAN.

RHEUMATOID ARTHRITIS (RA)

RA is a systemic chronic inflammatory disease of unknown etiology and is characterised by polyarthritis which may be progressive and permanently deforming and by extra-articular manifestations, such as rheumatoid nodules, pericarditis, and arteritis. Adult RA is commonly associated with a peculiar group of anti-IgG autoantibodies called rheumatoid factors which mainly belong to IgM or IgG classes, bind to the Fc portion of other IgG molecules, and form IgG-anti-IgG complexes in the circulation or joint fluid. RFs are detected in serum in up to 80% of adult patients with RA and often at high titer and in repeated tests. Nevertheless, RFs are not specific for RA and occur in other CTDs, such as Sjogren's syndrome, in chronic infectious diseases, such as infective endocarditis, tuberculosis, and hepatitis B and, although usually at low titer, in up to 20% of overtly normal elderly individuals.

Pathologically, RA is characterised by chronic inflammatory and proliferative changes in the synovial membrane (interior of joint capsule lined with synoviocytes), accompanied by inflammatory effusion in the joint space and destructive erosion of joint cartilage and adjacent bone cortex. Ultimately, the joint space may become obliterated and the bone ends united by fibrous or bony union (ankylosis).

In advanced RA, the synovial membrane becomes densely infiltrated with lymphocytes, plasma cells, and macrophages and extends over the articular surface as a membrane, called rheumatoid pannus, which erodes and replaces the underlying articular cartilage.

RA may also be accompanied by extra-articular lesions of which the rheumatoid nodule, a granuloma occurring in subcutaneous sites or elsewhere, is characteristic.

RFs in RA are produced not only in spleen and lymph nodes but also locally, and significantly, by plasma cells in the rheumatoid synovial membrane.

RF complexes formed with IgG locally, in synovial fluid, or in the circulation may activate C'-, neutrophil-, and macrophage-mediated mechanisms of joint and extra-articular injury and inflammation.

Cell-mediated immune mechanisms and proinflammatory cytokines are now a focus of interest in the study of the pathogenesis of RA. Briefly, T-lymphocytes, mainly T-helper/inducer cells and many of them activated, are the most abundant cells in the rheumatoid synovial membrane. Many activated macrophages, macrophage-like synoviocytes, and interdigitating reticular cells (strongly expressing HLA-DR antigen) are intermixed with the recruited T-cells and B-cells.

Among the several inflammatory cytokines released by the activated macrophages are TNF-alpha and IL-1 which are considered to be major mediators of joint inflammation in RA. Recent clinical approaches in the treatment of active RA include the use of new classes of immunomodulatory drugs that inhibit the action of these inflammatory cytokines, for example, that bind to and block TNF-alpha and its interaction with its cell surface receptor.

POLYARTERITIS NODOSA (PAN)

PAN, or simply polyarteritis, is a classical model of noninfectious systemic necrotizing vasculitis (a generic term) and involves small and medium-sized arteries of almost any organ (kidney, heart, intestine, skin) with the possible exception of lung. Pathologically, polyarteritis is characterised by segmental fibrinoid necrosis and polymorphonuclear leukocyte infiltration of all three layers of the vessel wall.

Involvement of the intima results in thrombotic occlusion and infarction of regions supplied by the affected artery. Destruction of the elastic lamina results in aneurysmal dilatations (nodosa) in some cases.

Significantly, up to 20-30% of patients with polyarteritis have hepatitis B virus antigenemia. HB surface antigen (HBsAg)- anti-HBsAg complexes are present in the circulation and deposited along with C' in the vascular lesions of such patients.

SYSTEMIC LUPUS ERYTHEMATOSUS (SLE)

SLE is a multisystem chronic inflammatory disease of unknown etiology which has a strong predilection for young women in the reproductive years, affects many organs, and is characterised by the production of multiple autoantibodies, typically antinuclear and anti-DNA antibodies. In some patients, autoantibodies are also produced against platelets, lymphocytes, and other cellular antigens.

The female: male ratio of SLE is about 9 to 1, and its prevalence in women is reported to be as high as 1:700. SLE is about a tenth as common as RA.

The clinical expression of SLE is extremely varied, but the most common findings include: fever, erythematous rash, arthritis, serositis (pleuritis, pericarditis), and nephritis. Skin photosensitivity, oral ulcers, hematologic disorders, and neurological abnormalities are also seen. SLE is a multisystem disease and mimics many other disorders.

Contributory factors may be genetic, hormonal, or environmental in nature. Genetic influence is suggested by family clusters of SLE, significant (approx. 25%) concordance of SLE in homozygous twins, and associations with HLA (DR2, DR3) and with inherited C' deficiencies. (Genetic studies of animal models indicate that multiple genes are involved in SLE-like disease expression). Hormonal influence is suggested by the 9:1 female:male ratio of SLE, the predilection for females during the reproductive years, and the frequency of onset in the early postpartum period. (In laboratory animal models, estrogens increase and androgens decrease the incidence of SLE-like disease). Environmental factors are indicated by drug-induced (iatrogenic) lupus syndromes caused by procainamide, hydralazine, and other drugs, and by skin sensitivity to sunlight exposure which may provoke the onset of SLE. The possible etiologic role of a specific infectious agent is speculative at this time.

Immunological Aspects of SLE

The discovery in 1948 of the "LE cell" was a major advance in diagnosis and led to the recognition that SLE and related CTDs are associated with the presence of antinuclear autoantibodies (ANAs). ANAs are a generic group of serum, mainly IgG, antibodies which react by indirect immunoflurescence test with acetone-fixed cell nuclei from a wide range of tissues and animal species (thus ANAs lack tissue and species specificity).

ANAs also react specifically with isolated and characterised nuclear antigens, such as deoxyribonucleoprotein, DNA in double stranded (ds) and single (ss) stranded configuration, and histones. Some ANAs also react with non-histone proteins, such as Sm antigen, various nuclear ribonucleoproteins (RNPs), nucleolar RNPs, and others. None of these reactivities has absolute specificity for any disease entity. The most specific for active SLE are anti-dsDNA antibodies which are rarely found at significant levels in other diseases.

Anti-DNA antibodies are involved in the pathogenesis of (type III) IC-initiated tissue injury and inflammation in SLE expressed as nephritis, vasculitis, and serositis. Autoantibodies are also produced in some lupus patients against platelets, erythrocytes, and T-cells, and these may cause antibody-mediated cell injury resulting in thrombocytopenia, hemolytic anemia, and lymphopenia.

Diagnosis of SLE

The diagnosis of SLE is based upon the appropriate clinical features and is confirmed by serological, pathological, and other laboratory findings. The fluorescent antinuclear antibody (FANA/ANA) test is the most widely used diagnostic screening test for SLE.

This test is positive, often in high titer, in >95% of patients with active untreated SLE, but positive reactions also occur in other CTDs, as previously noted, other conditions, and rarely in overtly healthy subjects, particularly family members of patients with SLE. While the ANA test has high sensitivity, it has low specificity for SLE. It is a useful diagnostic screening test in the evaluation of patients suspected of having SLE or a strongly ANA-related CTD or in ruling out these diseases.

Other Autoimmune Diseases

- Hashimoto's thyroiditis
- Myasthenia gravis

Infectious diseases

- Tuberculosis
- Histoplasmosis
- Infectious mononucleosis
- Chronic active hepatitis

Miscellaneous

- Pernicious anemia
- Ulcerative colitis
- Alcoholic cirrhosis of the liver
- Lymphoproliferative disorders

The pattern of nuclear fluorescence observed in the ANA test suggests the predominant type of antibody in the test serum and the nature of the reactive antigen. Anti-ds DNA antibodies typically give a rim pattern of nuclear fluorescence and are associated with SLE. Anti-histone antibodies typically give a homogeneous pattern and are associated with drug-induced LE and also SLE. Antibodies to nonhistone proteins and nuclear RNPs give a speckled pattern and are associated with SLE, Sjogren's syndrome, MCTD, and scleroderma. Antibodies to nucleolar RNPs are associated with scleroderma.

Other serological abnormalities of significance in SLE include: decreased C' activity; polyclonal hypergammaglobulinemia; cryoglobulinemia (cold precipitable globulins); and biological false positive STS, which is related to the presence of anti-phospholipid antibodies and associated with a syndrome of thrombosis and recurrent abortion.

Pathology of SLE

It is important to bear in mind that SLE is characterised by the total picture of clinical, immunological, and pathological features and that many of the pathological changes are nonspecific and sometimes may appear disproportionally small in comparison with the extent of clinical manifestations.

The principle pathological changes are seen in the skin, joints (nonerosive synovitis), kidneys (lupus nephritis), serous membranes (pleuritis, pericarditis), and heart (endocarditis, pericarditis).

The most typical histopathological lesions, aside from the presence of pathognomonic hematoxylin bodies, occur in blood vessels, kidneys, and skin.

In active SLE, vasculitis with subendothelial fibrinoid deposits may involve small arteries, arterioles, and capillaries of affected organs (kidneys, spleen, heart, lungs) and, rarely in the severest cases, acute necrotizing vasculitis may involve the entire vessel wall. Granular deposits of Igs and C' are seen in the acute vascular lesions by fluorescence microscopy.

In later stages, perivascular fibrosis occurs. The small penicillar arteries of splenic pulp typically develop concentric perivascular laminations of fibrous tissue to produce the typical "onion skin" lesion of spleen which is characteristic of, but no longer regarded as specific for, SLE.

The renal glomeruli are major targets of immune injury in SLE. Some 50% of SLE patients have glomerular disease (lupus nephritis) shown by urinalysis (hematuria, proteinuria, casts) or expressed clinically. About 70% of patients have glomerular disease indicated by light microscopy, and nearly all have glomerular abnormalities shown by immunofluorescence and electron microscopy.

There are several histopathological patterns of glomerular disease in lupus nephritis: mesangial, focal proliferative, diffuse proliferative, and membranous lupus nephritis.

This photomicrograph of diffuse proliferative lupus nephritis shows an increase in the size of the glomeruli and in the number of cells, reflecting the proliferation of endothelial, mesangial, and epithelial cells, including "cresentic" proliferation of parietal epithelial cells which are encroaching upon the urinary space. Additionally, GBM thickening is seen in peripheral capillary loops in a segment of the glomerulus at 9 o'clock. H&E.

The deposits contain IgG and protein components of the classical C' pathway, but IgM, IgA, proteins of the alternative C' pathway, and fibrin are sometimes present as well.

Electron microscopy shows the presence of electron dense deposits in corresponding locations in the mesangium and along the GBM beneath the vascular endothelium (subendothelial) or beneath the visceral epithelium (subepithelial). Subendothelial deposits when abundant increase the thickness and apparent rigidity of the GBM of peripheral capillary loops and produce the highly refractile "wire loop" lesion seen by light microccopy and characteristic of SLE.

There is strong evidence that C' activating DNA-anti-DNA complexes are involved in the pathogenesis of lupus nephritis, vasculitis, and serositis. Briefly summarised: circulating ICs, including DNA-anti-DNA complexes, are present in patients with SLE; high or changing levels of anti-dsDNA antibodies and decreased levels of serum C' correlate with disease activity; Igs, C', and, when studied, DNA antigen are deposited in the glomerular lesions; and Igs eluted from the glomeruli in vitro show a greatly increased concentration of anti-dsDNA (and anti-ssDNA) antibodies compared to serum. That is not to say that DNA-anti-DNA complexes are the only cause of IC-mediated mechanisms, because the antigen-antibody components of circulating ICs in SLE patients are not fully characterised.

DNA-anti-DNA complexes are also formed in situ in the glomeruli in SLE. DNA, presumably derived from the normal turnover of cells, is sometimes present in the plasma of patients with active SLE, and this circulating DNA has affinity for the GBM (apparently for type IV collagen). Circulating anti-DNA antibodies combine with this "planted" DNA to form finely granular deposits of Igs and C' along the GBM.

Skin lesions involving the face, trunk, and extremities occur frequently in SLE and take many forms: facial erythema, such as the classical "butterfly" rash distributed over the cheeks and base of the nose, urticaria, maculopapular lesions, ulceration, and alopecia. The presence of liquefaction degeneration of the basal layer of the epidermis and fibrinoid change at the epidermal-dermal junction is a microscopic characteristic of acute SLE. Immunofluorescence microscopy shows granular deposits of Igs and C' along the epidermal-dermal junction of acute lesions and also of uninvolved "normal" skin (lupus band test). The presence of these deposits in both involved and normal skin distinguishes SLE from other CTDs, such as scleroderma and dermatomyositis, and chronic discoid lupus (cutaneous LE) in which deposits occur only in the skin lesions. It is postulated that the deposits are ICs precipitated in situ as nuclear DNA released from damaged or senescent basal epidermal cells reacts with anti-DNA antibodies diffusing from dermal capillaries.

The joint involvement in SLE consists of nonerosive synovitis with an exudation of neutrophils and some lymphocytes but with the absence of joint destruction and deformity, such as seen in RA.

6

Pathogens to Animals

WAYS PATHOGENS SPREAD

In order to prevent the introduction of pathogens, one must understand how pathogens find their way onto backyard, hobby, and large corporate poultry or livestock operations. The spread of pathogens is primarily caused by the movement of animals, people, and contaminated equipment. Therefore, you can help prevent the spread of pathogens by controlling the movement of people and animals and avoiding contact with any potentially-contaminated equipment or objects.

CONTROLLING ANIMAL MOVEMENT AND CONTACT

The easiest and most frequent way to transmit a pathogen is for an infected animal to come into contact with a healthy animal. Most disease is transmitted nose-to-nose or by other direct contact with sick animals. New additions to flocks or herds are a very common way to introduce pathogens. Stray pets that wander onto your farm or land can be another source of pathogen transmission.

If a sick animal is allowed into a show or fair, it may infect other nearby animals. In addition to stray pets; wild animals, free-flying birds, and insects can be responsible for the introduction of a pathogen.

People Movement People, vehicles, equipment, and clothing that are in contact with an infected animal may be contaminated with pathogens. These contaminated articles include equipment, machinery, and other objects used for the transportation, care, and management of livestock and poultry. If pathogens can be spread from animal to animal at shows and fairs, does that mean that we cannot take an animal to the show? We can purchase animals and go to the fair, but you should isolate those animals when you return. The wisest decision is to choose shows and make livestock and poultry purchases from places where disease prevention is always given high priority.

Practical Steps that Improve Isolation

Isolation (a form of quarantine) prevents contact between animals within a controlled environment. The idea behind isolation is to prevent contact between healthy animals and an animal that is or could be infected with a pathogen. Animals new to the herd or flock should be isolated away from the rest of the herd or flock for at least four weeks. Always isolate sick animals and only return them to their original group when they've fully recovered. Livestock returning from fairs or shows should also be isolated, since they could have picked up a new pathogen at the show.

After returning to the farm, you should isolate your show animals to avoid the possibility of infecting other animals on your farm. Ideally, this should be in a completely separate place to avoid close contact. At minimum it could be a separate pen in a different building or at least a separate corner of the barn. This may represent some extra work, but it can be very important. Isolate all purchased animals for four weeks. If they are incubating a serious disease, the signs (i.e., diarrhea, hard breathing, etc.) of that disease will likely be observed in that 4 week isolation period.

This gives you some time to do follow-up testing or give booster vaccinations if needed. Always purchase good, healthy animals. Do not buy sickly animals, even if they are being sold for a "cheap" price. Livestock or poultry at bargain prices can be expensive!

Since people can carry pathogens on their body or clothing, it is necessary to control the visitors that may come in contact with your livestock or poultry. Always make sure visitors wear clean boots and clothing. It is best if visitors come only when you are there to escort them around your farm. Locked gates and doors will keep people out when you're not around. Signs and notices help to alert visitors of the potential risk that they may pose. Perimeter fences will keep people from accidentally wandering onto your property. Question visitors about the farms or animals they have visited or have been near recently, and ask if they have visited a foreign country within the past 5 days.

There are ways to help keep out wild animals, free-flying birds, and insects. Although it is impossible to totally prevent all contact between wildlife and our livestock, we can make barnyards and surroundings unattractive to many of these species. Cutting the grass and weeds around the outside of buildings will help prevent rodents from entering. Keep grain spills or other potential sources of food cleaned up and unavailable to wildlife.

Clean up old board piles or woodpiles and inspect buildings for possible hiding or denning areas. Inspect the haymow or other protected areas for evidence that cats, raccoons, or other animals that are likely using the hay or the straw for nesting areas. Store feed where wild animals, including rodents and birds, cannot make contact with it. Dispose of all waste feed in a way that will not attract pests. Maintain barns and buildings in good repair making it more difficult for animals or birds to gain access. Keep doors and windows shut when not needed for ventilation. Place screens or netting on the windows.

Location and Construction of Buildings

When planning the location of a barn or building used to house your livestock and/or poultry, isolation is key. The buildings or pens should be constructed to promote isolation from outside sources of disease. The facility should be a substantial distance from road traffic. It is important to consider possible exposure and the distance from other flocks, herds, or populations of domestic or wild animals. It is especially crucial that there is no nose-to-nose or beak-to-beak contact with other animals. There should not be any contact with drainage water or waste runoff from other animals. When planning the organization of either a large- or small-scale animal operation, the risk of disease can be greatly reduced whenever you keep different species and ages of the same species well separated. This is because pathogens producing little or no disease in one species can produce significant disease in another species. Mild infections in older animals can produce severe disease in young animals.Disease susceptibility within the same species varies with age.

Because they have acquired resistance, older animals may shed pathogens without getting sick themselves. These same pathogens can be serious to younger animals. Salmonella and E. coli are good examples of pathogens that can spread and produce extra trouble in young or newborn animals when they are not well separated from older stock.

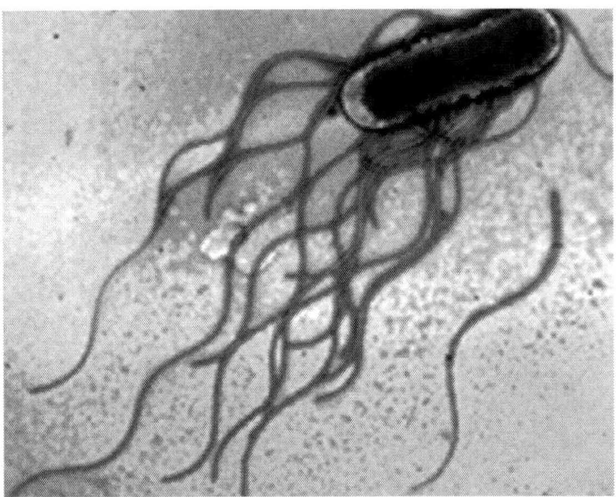

Fig. Salmonella magnified 3000 X

Various ways may be available to achieve species/age separation. They will be different depending on many factors such as the value of the animals, the economic and labour resources available to the herd/flock owner, and such other factors as farm size and marketing opportunities. They may include separate buildings, well-separated pens, different caretakers, or, ideally separate premises for different species and ages. Certain types of farm operations are able to ensure separation of ages by using the "all-in/all-out" management practice. All-in/all-out management allows only one age at a time on a premise. No new or younger animals are brought in until all of the older animals are removed and the barn has been cleaned. This allows for cleaning and disinfecting before a new group of animals is brought onto the premises. The farm environment can be a very comfortable place where pathogens can survive and multiply.

Animal buildings, barns and pastures, as well as litter and manure surfaces, are often damp environments favourable for a dangerous increase in pathogen populations. Depending on the number of animals, very large accumulations of manure and litter may develop and become areas where pathogens can survive and multiply for many weeks or months.

SANITATION

How can you avoid bringing your animals to pathogens? The best way is to maintain a clean, dry, and sanitary environment. Cleaning is very important in reducing pathogen levels. Cleanliness means freedom from dirt, filth, and debris. Although buildings and equipment can appear clean when they are not visibly dirty, many microscopic contaminating pathogens can continue to be present. They remain and survive by clinging to surfaces, and by hiding in tiny cracks and crevices. The use of a disinfectant kills these remaining pathogens

Proper disinfection results in reduction of pathogens. Reducing the amount of pathogens around your animals will decrease the risk of disease. Disinfectants are chemical agents that kill pathogens on contact. However, it is important to clean before disinfecting so that the pathogens become fully exposed to the disinfectant.

Pre-cleaning is doubly important because a disinfectant can be used up or inactivated by dirt. Disinfection and thorough, prompt drying are the last steps in the sanitation process. Without quick, thorough drying, any pathogens surviving the effects of the disinfectant can re-multiply to high numbers.

Hygienic Manure

Animal body wastes (feces, urine, etc.) ordinarily accumulate right in the environment (surroundings) where the animals spend most of their lives. In humans, of course, this does not occur because of toilets, plumbing and sewage treatment/disposal systems. Since flush toilets are not practical for animals, this ordinarily unavoidable problem for animals is partially reduced when manure or litter is managed in a way that, at least lowers the level of exposure to potential pathogens.

Accumulated manure must be managed to reduce its volume, its odor, and to kill pathogens and weed seeds. The essential parts of manure management include collection, transfer, and storage. Spreading manure onto fields is the most common method of disposal.

The sun will dry manure after it is spread onto fields which will kill many pathogens. A proper level of airflow over manure surfaces promotes drying and the decrease of many pathogen populations. In contrast, stagnant air contributes to the build- up of moisture, and an increase in pathogen populations.

Chemical treatments can be beneficial in manure management. Raising the pH to at least 12 for 30 minutes will kill most of the microorganisms present. Lime is usually used to raise the pH. Treatments that produce anaerobic conditions (very low oxygen) are also used. Anaerobic lagoons take advantage of a natural process where manure is digested by beneficial anaerobic bacteria. In contrast, aerobic lagoons add oxygen to the manure.

The addition of oxygen allows more common bacteria to survive and multiply and for the bacteria to break down the waste material. These bacteria convert the manure into carbon dioxide, water, and more beneficial lagoon bacteria. In animal pens or holding areas, we want the bedding and floor dry. Management of areas where water spillage or water accumulation is highest is very important. Good drainage and proper ventilation of buildings will help reduce the dangers of dampness. The frequent removal of wet litter and/or bedding and a continuous, modest flow of air over manure/litter surfaces will help to produce the drying needed to suppress bacteria.

THREATS TO ANIMAL HEALTH FROM THE VISIBLE TO THE INVISIBLE

Animals may be injured or die from the attacks by something as large and dangerous as a coyote or wolf to something just as dangerous, but only as small, as a molecule. This section takes you on a voyage into a world of animal health dangers.

This fascinating voyage ranges from dangers that are visible to the naked eye (predators), to some that are more easily seen with a hand lens (internal and external parasites), to some that can only be seen with a microscope (bacteria and viruses).

The voyage continues all the way on down to the sub-microscopic world of disease-producing molecules! More about these visible and invisible hazards to animal health and a short discussion about their prevention is presented below. The discussion is very basic and introductory.

We know that disease-causing pathogens that we cannot see threaten our livestock. In addition, there are also other threats that can be easily seen. Predators may easily be detected; however, many often appear at night or when people are not around. There are ways to help keep out predators that could potentially physically attack and harm your animals.

First, remove all easily accessible food supplies. This may be very difficult depending on the size of the farm and the amount of livestock feed on-hand. Keeping feed bins in good repair and sealed off will help prevent predators from entering areas where feed is stored. Keeping feed bins and feeding areas clean is very important.

An accumulation of waste feed in and around feed bunks attracts animals. A second suggestion is to remove water supplies. This is even more difficult on most farms. Stock tanks often provide a constant water supply and are attractive to predators such as raccoons. Preventing any unnecessary pools of water will help.

A third suggestion is to modify habitat and reduce access. Clearing brush and keeping weeds away from barns and buildings will help deter animals just like it will help deter rodents. Finally, you can trap or control predators. There are a variety of traps available to catch animals.

Barns and buildings should be kept in good repair. Predators often enter barns though open doors, or cracks and holes that may exist. Closing doors and patching cracks and holes helps to reduce the problem. Wire mesh or screening on windows can also help.

EXTERNAL PARASITES

Small yet visible threats to livestock include external parasites such as ticks, flies, fleas, lice, mosquitoes, mites, grubs, etc. Parasites are a threat to livestock health just as microbial (invisible) pathogens are a threat. Parasites can transmit and spread microbial pathogens in addition to the harm and damage that they naturally cause by irritating animals and sapping their energy. The most common way to control external parasites is through the use of pesticides or insecticides. Pesticides kill and help control the parasite populations. Pesticides are applied as sprays, dips, pour-ons, dusts, injectables, pastes, boluses, etc. Ear tags impregnated with insecticide are commonly used in cattle.

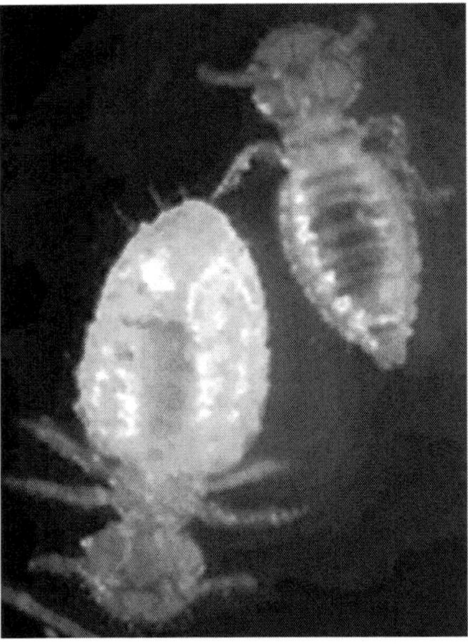

Fig. Lice

Pesticides are very effective however, pesticides alone will not control a threat such as flies. Keeping the environment clean and sanitary helps eliminate fly breeding areas. Manure management will help reduce fly breeding areas. Fly eggs and larvae in thinly spread manure are killed by drying and heat.

INTERNAL PARASITES

Livestock and poultry can be attacked by a variety of large and small worm-like parasites such as roundworms and tapeworms. These internal parasites mainly invade the digestive passages while some also infest an animal's breathing passages. Other parasites can go even deeper into the "donut" or animal body reaching various vital organs and body tissues. Most parasites enter an animal's body when the animal eats the egg (ova) or an early life stage of the parasite. These ova or intermediate life stages come from the adult parasite reproducing and living inside an animal. They are typically passed by way of animal droppings onto the ground, or into the bedding or litter. As a result, an effective preventive strategy for many internal parasites rests on keeping animals from eating feed or licking surfaces contaminated by animal waste. It is a wise practice to keep animal yards, pens and buildings or other concentration points as clean from urine and accumulated fecal material as is practicable.

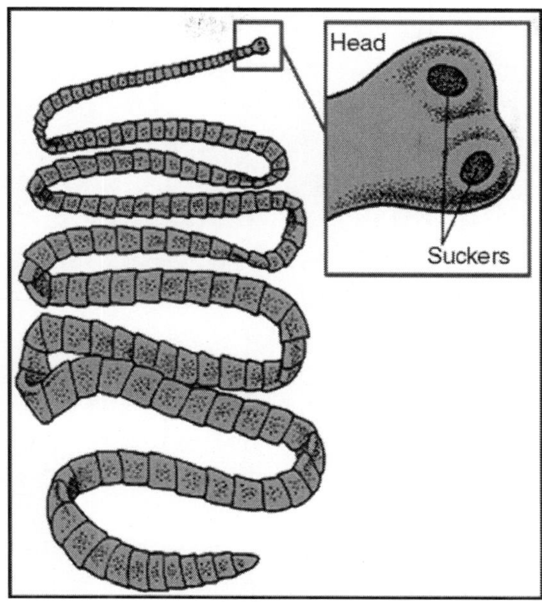

Fig. Tapeworm

Various powerful chemicals are sometimes used to treat different types of infestation. Their improper or inappropriate use may produce more damage to your animals than would be done by the parasites alone. Consequently, it is best to seek guidance from your local veterinarian before you treat for worms. In the final analysis, prevention is often less expensive than treatment.

Bacteria and Viruses

Bacteria and viruses are visible only when they are magnified hundreds to many thousands of times. They are not only able to attack the skin and digestive tract or respiratory linings of an animal's body, but, are often able to devastate the entire body including such organs as the brain, heart, liver and spleen. Their extremely small size makes it possible for these pathogens to survive long periods of times outside an animal's body. They can survive in fur, hair and feathers, in nasal and other discharges from a sick animal, and in animal urine and fecal droppings.

Viruses survive, and bacteria can actually multiply, in animal bedding, litter, and manure. Many survive long periods of time in the tiny particles of dust or soil present in the farm environment. With many tiny, scattered hiding places they can easily be carried to a new farm or group of animals riding on a person's clothing, on the surfaces of boxes, crates or equipment, and on the wheels of cars and trucks. Bacteria are single-celled microorganisms. They may live free in the environment or within a living cell. E. coli (Escherichia coli), Streptococcus, Staphylococcus, and Salmonella are a few examples of bacteria that can cause disease.

Viruses are tiny organisms that only grow inside the cells composing the animal's body. All viruses rely on a live animal host to reproduce. Examples of viruses that affect humans include the common cold (rhinovirus) or the flu (influenza) virus. Viruses can infect animals and cause respiratory symptoms (influenza viruses), diarrhea (rotavirus and coronaviruses), and numerous other disorders.

Although common viral or bacterial pathogens are unwelcome visitors or residents, certain ones can be especially troublesome. These special pathogens produce unusually high losses and/or unusual behaviour such as extensive tenderfootedness, incoordination, or slobbering. Milk or egg production may cease or drop sharply. Your daily activities in caring for and feeding your animals put you in an excellent position where you could be among the first to spot the possible emergence of an especially unwelcome pathogen.

As previously emphasized, isolation, traffic control, hygiene, and sanitation are the main ways to keep these essentially invisible pathogens from spreading to, and from reaching levels that can infect your animals! Medications, antibiotics, and vaccines are additional measures available to minimize the effects of bacterial or viral infections. Although powerful allies in an animal's battle with one or more pathogens, they are really the second line of defence.

The first line consists of all the steps you take to keep these microbes out or their numbers down to begin with. Again, those primary steps are herd/flock isolation, traffic control, hygiene, and sanitation.

Dangerous Molecules Prion

A prion is a very unusual infectious agent that is capable of causing an infection or disease. It is believed to be a self-reproducing protein structure that is similar to a virus. Prions cause prion diseases such as transmissible spongiform encephalopathies (TSE's). Unlike other infectious agents, prions produce no body-defending immune response.

Mold Toxins. Mold growth takes place in feeds when they are damp. Mold does not always mean that you cannot use the feed. Most molds are not toxic when fed to livestock; however, some are very toxic. The toxic by-products produced by molds (fungi) are known as mycotoxins (mold toxins). Mold toxins are very diverse because they are produced by many different molds.

Fungi that have the ability to produce mycotoxins are very common in grain and other livestock feeds and in the facilities and equipment used for transportation and storage of feed. They can be very dangerous causing reduced growth rates, lowered immunities, and increased susceptibility to various infectious diseases. Many feed companies are working on ways to sample and test feeds and grains for mold toxins before their sale.

Animals feeds must be kept dry. We must realise that feeding any moldy feedstuffs is a risk to the health of our animal. There is less risk in feeding moldy feeds to fattening animals than for lactating or pregnant animals. The risk is also different among different species of animals.

Dryness is the simplest way to control the growth of mold. Take the following steps when evidence of mold growth (musty odor, visible mold) is suspected.

- Stop the moisture source (fix the leak) correct condensation problems in outside storage bins.
- Thoroughly dry or discard all porous items.
- Scrub mold off hard surfaces with detergent and water, and thoroughly dry. Chlorine-containing products are of help in killing mold spores.

Bacterial Toxins

Many bacterialpathogens have the ability to produce toxins which add to the disease they produce, thus making them a double threat to your animals. Some toxin-producing bacteria form their poisons outside the body (botulism), while others do so inside the body (tetanus, black leg, and gangrene). These toxic bacteria can present a much greater risk than many ordinary bacteria. Vaccines are used to aid in their prevention, particularly where risks of exposure or infection are high.

SAFEGUARDING ANIMAL HEALTH

The position of a flock or herd is not unlike that of a populous city of which the public health largely depends on the functions of an intelligent "health officer." The owner, above all, should function as a health officer to his/her flock or herd. This element of animal management completes the picture of what it takes to keep animals safe and healthy. The health officer idea is especially important to many young people. After all, in the next few years, they may become the owners or managers of even larger populations of animals.

They need to be alert to threats to the health of their own animals, and to those belonging to others. They should be quick to recognize and properly respond to unusual diseases and to oversee and manage their own operations to prevent or minimize the effects of common diseases. The search for and application of disease prevention knowledge can be an enjoyable, enriching lifelong experience. An alert, active disease prevention "mind set" provides many dividends! By protecting your own animals from pathogens you are also helping to protect the animals of others.

CONTROL OF ANIMAL DISEASE

The control of animal diseases and the promotion and protection of animal health are essential components of any effective animal breeding and production programme. Despite remarkable technical advances in the diagnosis, prevention and control of animal diseases, the condition of animal health throughout the developing world remains generally poor, causing substantial economic losses and hindering any improvement in livestock productivity.

In developing countries, animal health services were established with the main objective of controlling major contagious and infectious diseases, such as foot-and-mouth disease, rinderpest and contagious pleuropneumonia, as well as parasitic diseases, such as trypanosomiasis and tick-borne diseases. This was obviously the first priority, since the control of these diseases is a prerequisite to any successful livestock development programme.

With the present concern for sustainable economic development, more attention is now being given to other diseases that affect livestock productivity, such as helminthiasis, nutritional diseases, reproductive disorders, etc.

The successful control of disease depends initially on its timely and accurate recognition and on the presence of sound diagnostic capabilities based on effective working links between laboratories and field services. Emergencies created by outbreaks of major infectious diseases demonstrate the need for establishing, strengthening and improving such diagnostic services. As well, particular attention should be given to the development of an efficient animal disease information system.

Beyond the national level, a general increase in the movement of animals and animal products underlines the importance of international cooperation in the prevention and control of animal diseases.

Most animal health services in developing countries do not have at present adequate technical and administrative infrastructures to carry out the tasks and duties necessary for the efficient control of animal diseases and for consumer protection.

In many developing countries, there is either a shortage of skilled veterinary personnel or their services are not correctly utilized. The problem is exacerbated by deficient veterinary infrastructures and inadequate disease control programmes, veterinary legislation and information services, as well as a lack of transport, communications, veterinary products and equipment.

The most common problem is the shortage of funds to sustain the activities of veterinary staff. In some developing countries, the animal health services are not given the appropriate legal power in the administrative system.

These shortages significantly reduce the effectiveness of animal health services control measures against major animal diseases. By exploiting existing conditions, local resources and international help, possibilities for improving of animal health service programmes are increased.

Veterinary education and training should receive the highest priority; however, more emphasis must be placed on the qualitative and practical aspects. Effective personnel development requires adequate planning of staff requirements, improved curriculums for undergraduate and postgraduate studies, training of auxiliary personnel and greater interregional cooperation to exploit existing training facilities fully, among other things.

Animal health services should be further developed by improving their efficiency. While the control and prevention of major infectious diseases clearly remains a government responsibility, some veterinary tasks such as the treatment of individual animals could be undertaken in other ways. Privatization is one way of improving some sectors of animal health and of responding suitably to the needs of animal owners.

Other ways include contracting out certain services, creating farmer cooperatives and producer associations, recovering government costs more efficiently and using the revenue thus generated for selective subsidies. Several countries have already taken steps to reorganize their veterinary services in this way.

Developing countries must address this problem. This publication aims to assist them in improving livestock production through better control of major animal diseases.

It is not intended to be a comprehensive text describing in-depth all aspects of a complex subject with worldwide variations. Instead, it is meant to serve as a guideline, providing general background information on basic topics. Its main objective is to assist animal health authorities in their organization, planning and management activities.

The contents cover major problems facing official animal health services in developing countries in contributing to the production of food of animal origin and livestock development as integral components of general social, economic and agricultural development.

Other priorities of animal health services include the protection of humans against diseases that may be transmitted by animals and the production of safe food.

Biological and pharmaceutical production may be the responsibility of some services, however, only aspects of the control and management of veterinary biologicals and drugs are included in this publication.

The major issues dealt with in this publication are the objectives, functions, organization and management of animal health services. Without going into detail, relevant information, statements and recommendations make up the basic text. This general approach does not permit specific disease control or specific social, economic and ecological conditions to be dealt with. However, information is provided in the annexes on specific topics and in the selected bibliographies of general interest publications.

7

Musculatory Skeletal System

This system consists of the bones and the muscles (meat).The bones form the skeleton which is the framework within the body. It carries weight and supports the body. Bones are connected together so they can move. The places where this happens are called joints. The bones are held together at the joints by elastic strands called ligaments. Between the bones is a softer material called cartilage (gristle) which cushions the bones at the joints when the body moves. Bones are very hard and contain minerals. Each bone has a name such as the scapula (shoulder blade) and skull (head). There are about 200 bones in the body.

Muscles are joined at both ends to the bones. The muscles are the meat of the body and when they contract (shorten) or relax (lengthen) they make the bones move. Mammals are vertebrates by definition, this means that all mammals have an internal bony support structure to which muscles and ligaments are attached. This is what we call a skeleton.

The basic plan of the mammalian skeleton, as seen above, is fairly straight forward. It consists of a head at one end of a vertebral column from which extend ribs to support the working organs and four limbs for locomotion. The vertebral column ends in a tail. The huge range of lifestyles and habitats utilised by mammals means that a great deal of variety exists between different groups.

Some species lack a tail, others lack apparent hind limbs and the skull is very variable. Here we are giving a general outline of the mammalian skeleton with some notes to indicate particular variations. To understand the skeleton fully you should also have a look at the bones and joints.

VERTEBRATE SKULL IN GENERAL AND THE MAMMAL SKULL

The vertebrate skull in general and the mammal skull in particular is a complex amalgam of bones, not just one or two but about 34 bones, if they all are present, make up the skull and lower jaw. Most of these bones are now fused together. Anyone who has looked carefully at a mammal skull wmust have noticed the crinkly lines where the once individual bones meet. Most of them come in pairs, one each side. The skull can be divided into 3 basic parts; the Braincase (enclosing the brain); the Rostrum (the snout and upper jaw); the Lower Jaw.

Some of the main bones of the skull are:

- Nasal bones - the roof of the nasal cavity
- Maxillary bones - the main bones of the upper jaw, including the Zygomatic arch; the bones to which the strong jaw muscles are attached and which form the boundaries of the eyes. The Zygomatic arches also contain the Jugal bones and the Squamosal bones.
- Frontal bones - the top of the skull in most mammals. These are the bones that horns and antlers grow from.
- Parietal bones - these form the roof and back of the Braincase.
- Occipital bones - the lower back of the braincase (actually 3 bones fused together; basioccipital, exoccipitals and the supraoccipital).
- Dentary bones - the lower jaw.

THE VERTEBRAE - SPINAL COLUMN

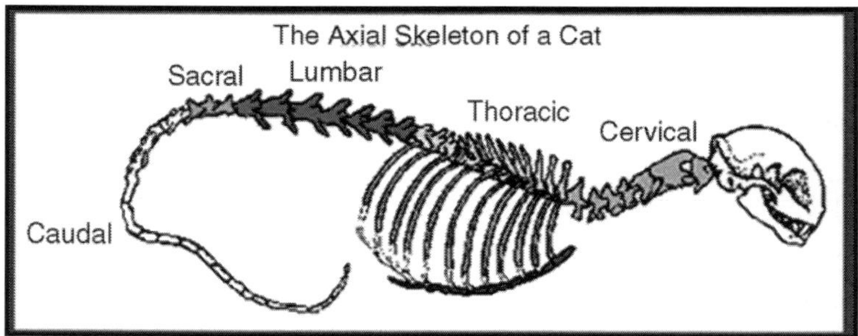

The spinal column in mammals, as in other vertebrates, is composed of a series of small bones with one or more holes through their centre. These holes are aligned to make a highly protected tube. This tube houses and protects the spinal chord a thick sheath of nerves and ganglia that is the unifying characteristics of the Phylum Chordata. The Phylum Chordata contains the Subphylum Vertebrata and hence the Class Mammalia.

In many mammals the vertebrae can be seen to be divided into five distinct regions, though in some groups such as the whales they are pretty indistinct. Where the adjacent vertebrae meet each other they have special smooth, flatish surfaces called zygopophyses. These contact surfaces are also protected by cartilage. Cervical Vertebrae - the neck region of an animal is supported by the cervical vertebrae and normally in mammals there are 7 of them. The first vertebra, the one immediately behind the skull is called the 'Atlas'. It has two large depressions in its front face which accept (articulate with) the occipital condyles, two bumps unique to mammals found at the base of the skull. It also has a slot in the rear face to accept the odontoid process, a forward project of bone on the second vertebra. This second vertebra is called the 'Axis'.

The rest of the cervical vertebrae do not have special names. They are often cemented together for extra strength in digging and swimming mammals. Thoracic Vertebrae - these are the bones from which the rib bones extend. They often have large dorsal spines. These are usually 12-15 thoracic vertebrae.

The spines help support the muscles that lift and control the neck and head. Lumbar Vertebrae - the third part of vertebrae are the lumbar vertebrae. These are the rest of the spine down to where the back legs connect. Normally there are only 6 or 7 of them except in the toothed whales (Odontoceti) where there can be as many as 20. Lumbar vertebrae often have numerous spines and processes and can look quite complicated.

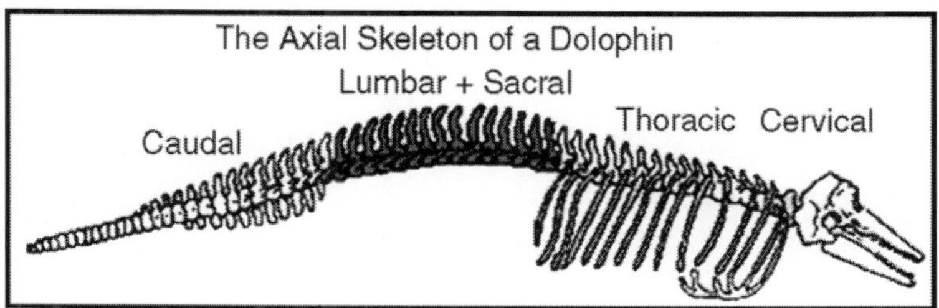

Sacral Vertebrae - these bones support the pelvic girdle and are often fused together. There are usually 3-5 in number but can be more, up to 10 in the Edentata for instance. Caudal Vertebrae - the last lot of vertebrae are the bones of the tail.

They are generally smaller and less complicated than the rest of the vertebrae. They also do not contain the spinal column which ends at or before the Sacral vertebrae, though they do contain some nerves and blood vessels. In mankind and Chimpanzees, the Caudal vertebrae are reduced to 4 in number and are fused to form the Coccyx.

Rib bones - these support the vital organs, making a cage within which heart and lungs can be carried in protection. Ribs normally arise from the thoracic vertebrae but in a few groups (Edentata, Monotremata) some species have ribs arising from the cervical vertebrae. Ribs end under or in front of the animal along each side of the sternum (breast bone). In many species the lower or posterior ribs do not meet the sternum, these are called floating ribs. In most groups articulated ribs (those which meet and join with the sternum) outnumber the floating ribs easily, but in Whales and Dolphins it is the other way around partly because of a reduced sternum (meeting in only 6 pairs of ribs as compared to 9 pairs in a cat) and partly because of a greater number of ribs; 15 pairs versus 12 pairs in a cat. The Sternum or breastbone is actually made up of a number of smaller bones called 'sternebrae'.

The front part of the sternum is keeled (to give a larger area for muscle attachment) in bats who need to flap their wings, and in many digging species who also need large muscles to work their forelimbs, *i.e.* moles. Most of the remaining bones are to do with limbs, either being part of them or supporting them.

The two supporting areas are called girdles (Pelvic girdle and Pectoral girdle). The Pelvic girdle consists of two halves each of which is made up of three bones fused together.

These two halves are called the 'innominate bones'. The three pairs of bones which make up each innominate bone are the Ilium, the Ischium and the Pubic bones.

In marsupials and monotremes another pair of bones are present called the Epipubic bones. The exact role that these play in the life of the Marsupials and Monotremes which possess them is unknown.

The pectoral girdle consists of two pairs of bones, the large flat Scapula (shoulder bone) and the much smaller and more slender Clavicle (collar bone). The clavicle runs from the far end of the scapular to the sternum in most mammals, though in the Monotremata it meets the interclavicle instead.

Monotremes also have a pair of bones called the Epicoracoid or Precoracoid bones. In some mammals such as the Anisodactyla, Perissodactyla, Mysticeti and Odontoceti (Horses, Pigs, Deer, Buffaloes, etc. and Whales) the clavicle is absent. The scapula is the bone that the legs or arms start from, *i.e.* the humerus joins the far end of the scapula at the glenoid fossa.

BASIC VERTEBRATE LIMBS

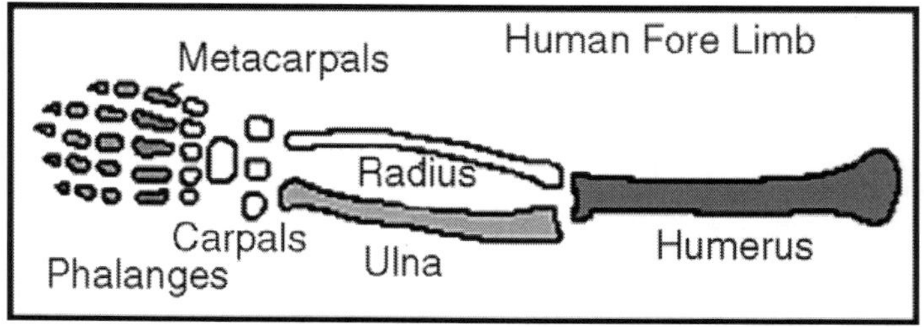

The basic vertebrate limb evolved as the first amphibian like fish left the warm seas for the new dry land. It is called the pentadactyl limb, from the Greek penta for five because it ends in five digits or phalanges.

This business of having five separate endings to a single limb was quite an important new development when it first occurred. It was obviously a good thing because it has been very successful. The pentadactyl limb supported the dinosaurs for millions of years and has played a major role in mankind's technological development.

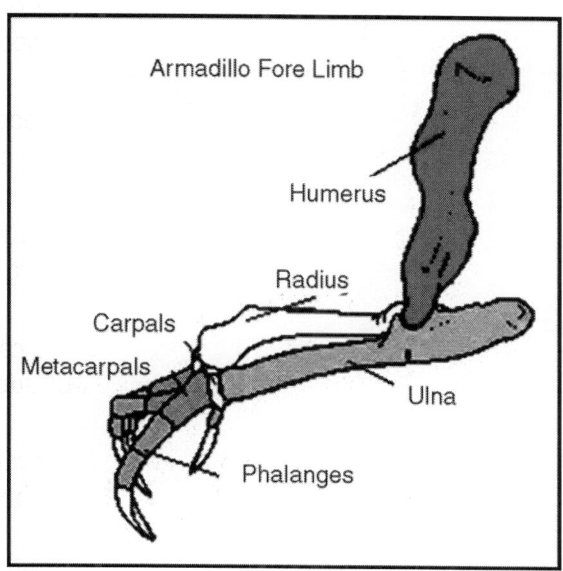

It remains the same all across the vertebrate classes. In mammals it perhaps reaches its peak and greatest diversity. Though the fore and hind limbs are basically the same they have different names for the various bones.

Thus starting at the body and moving outwards the forelimbs consist of:

Humerus	Ulna + Radius	Carpel bones	Metacarpel bones	Phalanges or Digits
Upper arm	Fore arm	Wrist	Palm of hand	Fingers

Hind limbs consist of:

Femur	Patella	Fibula + Tibia	Tarsal bones	Metatarsal bones	Phalanges or Digits
Thigh	Knee	Shin	Ankle bones	Foot bones	Toes

The lower part of both limbs consists of a pair of bones running parallel to each other. The Forearm consisting of the Ulna and Radius and the Shin consisting of the Fibula and Tibia. This arrangement allows for greater flexibility and stronger turning/twisting movements than a single bone would.

Though this is not terrible important in some mammals which only use their limbs for walking on, for those who use their limbs, prticularly the fore limbs, for grasping, climbing and or digging it is of great value and has contributed greatly to the diversity of mammals.

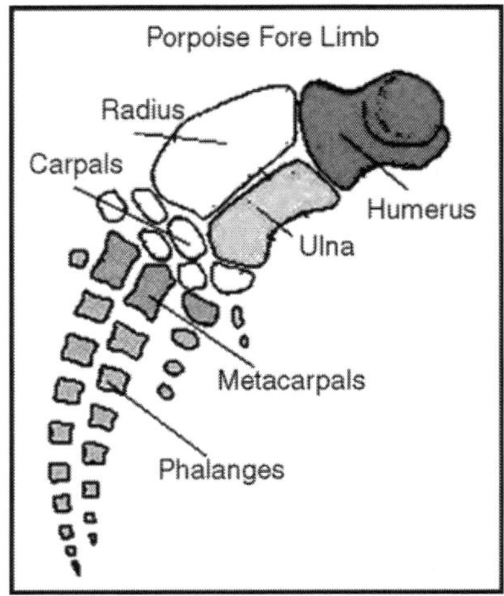

The Carpels and or Tarsals come in 3 rows see the Human fore limb above and make up the wrist. However as can be seen from the other fore limb images their number is often reduced and their role less important than it is in the classic design.

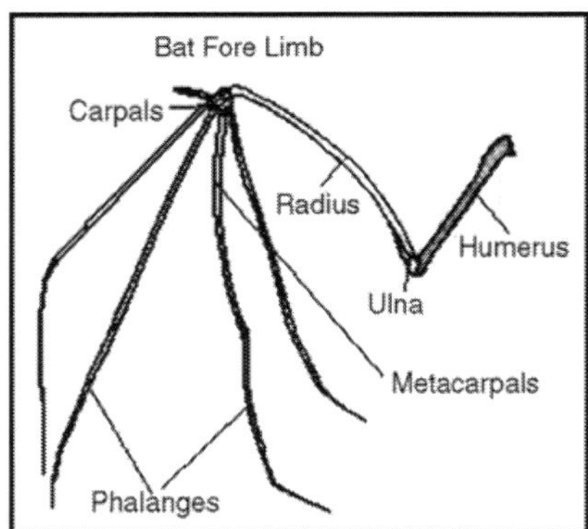

These various other limb bones also have varying degrees of importance in different families of mammals. They have all been modified by the forces of evolution to fulfil different roles relevant to the different way various animals use their limbs.

In Whales and Dolphins, for instance the number of finger bones 'phalanges' in the forelimb has greatly increased. This is known as hyperphalangy. In some families not all the bones mentioned above still exist, some have been lost in the course of evolution. Like all bones, limb bones contain various 'processes' (special extensions of bones to aid muscle attachment) and fossae (smooth surfaces where two bones meet and articulate [move in relationship to each other]).

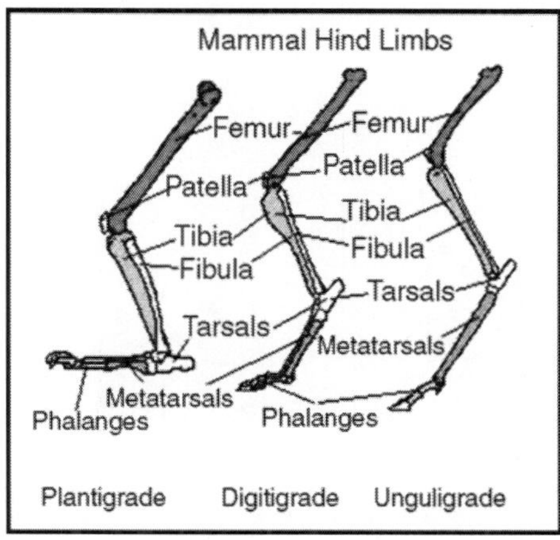

8

Circulation in Animals

Materials formed in one part of the body have to be taken up to other parts where they are needed or to be got rid of. This is an essential requirement of most animals. This function is performed by the body fluids.

DISEASES OF THE HEART AND CARDIOVASCULAR SYSTEM

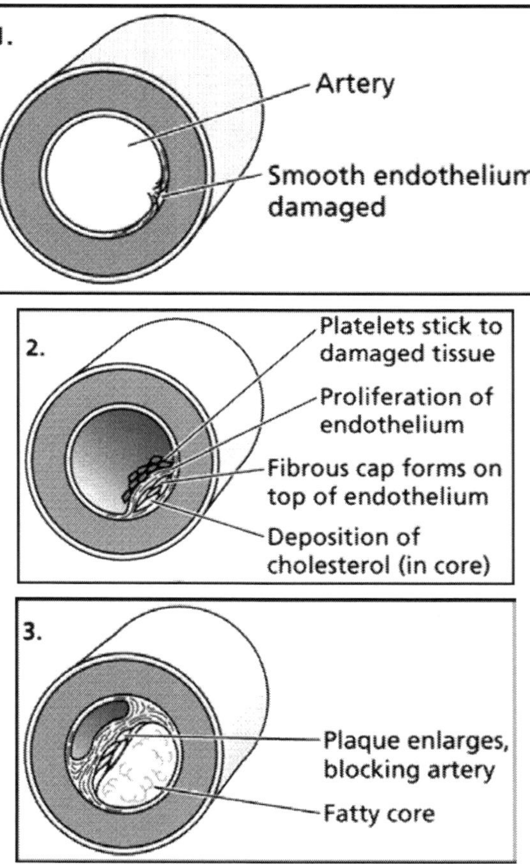

Fig. Development of Arterial Plaque.

Cardiac muscle cells are serviced by a system of coronary arteries. During exercise the flow through these arteries is up to five times normal flow. Blocked flow in coronary arteries can result in death of heart muscle, leading to a heart attack.

Blockage of coronary arteries, is usually the result of gradual buildup of lipids and cholesterol in the inner wall of the coronary artery. Occasional chest pain, angina pectoralis, can result during periods of stress or physical exertion. Angina indicates oxygen demands are greater than capacity to deliver it and that a heart attack may occur in the future. Heart muscle cells that die are not replaced since heart muscle cells do not divide. Heart disease and coronary artery disease are the leading causes of death in the United States.

Hypertension, high blood pressure (the silent killer), occurs when blood pressure is consistently above 140/90. Causes in most cases are unknown, although stress, obesity, high salt intake, and smoking can add to a genetic predisposition. Luckily, when diagnosed, the condition is usually treatable with medicines and diet/exercise.

The Vascular System

Two main routes for circulation are the pulmonary (to and from the lungs) and the systemic (to and from the body). Pulmonary arteries carry blood from the heart to the lungs.

In the lungs gas exchange occurs. Pulmonary veins carry blood from lungs to heart.

The aorta is the main artery of systemic circuit. The vena cavae are the main veins of the systemic circuit. Coronary arteries deliver oxygenated blood, food, etc. to the heart. Animals often have a portal system, which begins and ends in capillaries, such as between the digestive tract and the liver.

Fish pump blood from the heart to their gills, where gas exchange occurs, and then on to the rest of the body. Mammals pump blood to the lungs for gas exchange, then back to the heart for pumping out to the systemic circulation. Blood flows in only one direction.

Component of Blood

Plasma is the liquid component of the blood. Mammalian blood consists of a liquid (plasma) and a number of cellular and cell fragment components. Plasma is about 60 per cent of a volume of blood; cells and fragments are 40 per cent.

Plasma has 90 per cent water and 10 per cent dissolved materials including proteins, glucose, ions, hormones, and gases. It acts as a buffer, maintaining pH near 7.4. Plasma contains nutrients, wastes, salts, proteins, etc. Proteins in the blood aid in transport of large molecules such as cholesterol.

Red blood cells, also known as erythrocytes, are flattened, doubly concave cells about 7 μm in diameter that carry oxygen associated in the cell's hemoglobin. Mature erythrocytes lack a nucleus. They are small, 4 to 6 million cells per cubic millimetre of blood, and have 200 million hemoglobin molecules per cell. Humans have a total of 25 trillion red blood cells (about 1/3 of all the cells in the body).

Red blood cells are continuously manufactured in red marrow of long bones, ribs, skull, and vertebrae. Life-span of an erythrocyte is only 120 days, after which they are destroyed in liver and spleen. Iron from hemoglobin is recovered and reused by red marrow.

The liver degrades the heme units and secretes them as pigment in the bile, responsible for the colour of feces. Each second two million red blood cells are produced to replace those thus taken out of circulation.

White blood cells, also known as leukocytes, are larger than erythrocytes, have a nucleus, and lack hemoglobin. They function in the cellular immune response.

White blood cells (leukocytes) are less than 1 per cent of the blood's volume. They are made from stem cells in bone marrow. There are five types of leukocytes, important components of the immune system.

Neutrophils enter the tissue fluid by squeezing through capillary walls and phagocytozing foreign substances. Macrophages release white blood cell growth factors, causing a population increase for white blood cells. Lymphocytes fight infection. T-cells attack cells containing viruses. B-cells produce antibodies.

Antigen-antibody complexes are phagocytised by a macrophage. White blood cells can squeeze through pores in the capillaries and fight infectious diseases in interstitial areas.

Platelets result from cell fragmentation and are involved with clotting. Platelets are cell fragments that bud off megakaryocytes in bone marrow. They carry chemicals essential to blood clotting. Platelets survive for 10 days before being removed by the liver and spleen. There are 150,000 to 300,000 platelets in each millilitre of blood.

Platelets stick and adhere to tears in blood vessels; they also release clotting factors. A hemophiliac's blood cannot clot. Providing correct proteins (clotting factors) has been a common method of treating hemophiliacs. It has also led to HIV transmission due to the use of transfusions and use of contaminated blood products.

The Lymphatic System

Water and plasma are forced from the capillaries into intracellular spaces. This interstitial fluid transports materials between cells. Most of this fluid is collected in the capillaries of a secondary circulatory system, the lymphatic system. Fluid in this system is known as lymph.

Lymph flows from small lymph capillaries into lymph vessels that are similar to veins in having valves that prevent backflow. Lymph vessels connect to lymph nodes, lymph organs, or to the cardiovascular system at the thoracic duct and right lymphatic duct.

Lymph nodes are small irregularly shaped masses through which lymph vessels flow. Clusters of nodes occur in the armpits, groin, and neck. Cells of the immune system line channels through the nodes and attack bacteria and viruses travelling in the lymph.

BASICS OF CIRCULATORY SYSTEMS: OPEN VS. CLOSED

Animals are comprised of trillions of cells, and each cell in the body needs a supply of oxygen and nutrients in order to survive. To deliver these things and take away wastes, nearly all animals need a special transit system which is called the circulatory system.

There are two types of circulatory systems:

- Open and
- Closed.

The open circulatory system is found in arthropods and mollusks. The body cavity of these animals is called a hemocoel. The heart pumps a blood-like substance directly into the hemocoel where it moves freely and the organs are able to directly access the nutrients. In the hemocoel, blood and interstitial fluid (a special fluid which surrounds cells) is combined into a substance called hemolymph. This is the goo which sometimes squishes out of bugs when they get stepped on.

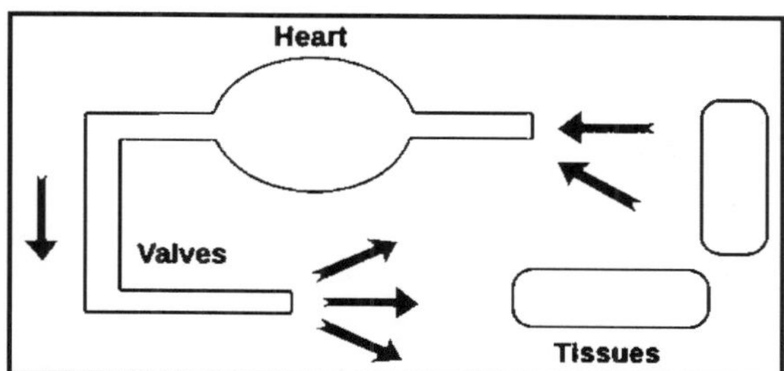

Fig. A Diagram of an Open Circulatory System. The Heart Pumps Hemolymph out into the Body Cavity Shaded in Pink Where it can Directly Bathe the Tissues and Provide Nutrients and Oxygen to them without the use of Blood Vessels.

A disadvantage to this system is that the fluid cannot be sent to a particular region of the body. Once it is pumped into the body cavity, it can be shifted when the animal moves, but precise delivery is difficult.

While the open system works fine for smaller animals, it would not be efficient in larger animals because it would require a tremendous amount of work to pump blood through a single large open body cavity. Imagine a pool filter that has to circulate every drop of pool water within a couple of minutes. Yikes.

Many large animals utilise a closed circulatory system. This means that the blood is enclosed in a series of vessels rather than flowing freely inside the body cavity. By retaining the blood inside tiny blood vessels it can be pumped to the far reaches of larger animals. It also gives the animal the possibility of pumping more blood to some areas and not others. A disadvantage of the closed system is that blood is no longer in direct contact with the plasma membrane of the cells. Diffusion of nutrients must first take place in two steps, first out of the blood vessel and then into the cell.

CIRCULATORY SYSTEM COMPONENTS

HEART

The heart is the key to any circulatory system because it provides the muscle power to move the blood throughout the entire body, even to itself. The arteries that supply blood to the heart are called coronary arteries, and they are the bright red blood vessels shown crisscrossing the top of the heart in the picture. Hearts come in many sizes; typically the size of the heart is directly proportional to the size of the animal. The heart of a tiny fairy wasp is less than 0.2mm. Otherwise, its heart would take over the full wasp length. The heaviest heart ever recorded belonged to a great blue whale and weighed 1,980 pounds. That's a ton of heart.

Apart from size, not all hearts have the same structure, either. An insect's heart consists of a tube-like structure which runs along an insect's back and contracts to move hemolymph from the posterior to the anterior. In the abdominal section there are a series of valves, called ostia, which allow hemolymph to enter the heart and get pumped into the anterior of the hemocoel. Many other animals have hearts with several distinct chambers. Fish have a two-chambered heart. The two cavities are called the atrium and the ventricle. Blood always enters the heart in an atrium (just like an entrance to a building) and always leaves the heart through a ventricle.

For this reason, ventricles are always stronger than atriums because they provide the golden push that sends the blood on the long journey throughout the body. If atria and ventricles were Matthias Schlitte's arms, the atria would be the left and the ventricles would be his right. Amphibians and reptiles have a three-chambered heart with two atria and one ventricle. Birds and mammals have a four-chambered heart with two atria and two ventricles. Between the heart chambers, there are heart valves, which are flaps of tissue that prevent backward blood flow. We will discuss the pattern of blood flow through the heart later.

As a muscle, a heart contraction is stimulated by an electrical signal created by a group of cells in an area of the heart called the sinoatrial node which tells the heart to beat every 1.0-1.6 seconds. A damaged sinoatrial node can be replaced with an artificial pacemaker that tells the heart when to beat. A pacemaker can be seen on the X-ray image shown below. Pacemakers can last for several years until, like all electronics devices, the batteries wear out.

Blood Vessels

The system of blood vessels consists of arteries, veins, and capillaries. Arteries carry blood away from the heart. They are composed of three layers, and the inside layer is coated with special anti-clotting factors which keep the blood from clogging. Unlike a clogged toilet, a clogged blood vessel can actually kill you. It's important to keep blood flowing smoothly. The middle layer contains muscle and elastic fibres, and the outside is comprised of connective tissue and elastic fibres. The elasticity of an artery allows it to flex and accommodate blood being pumped into it with a lot of force.

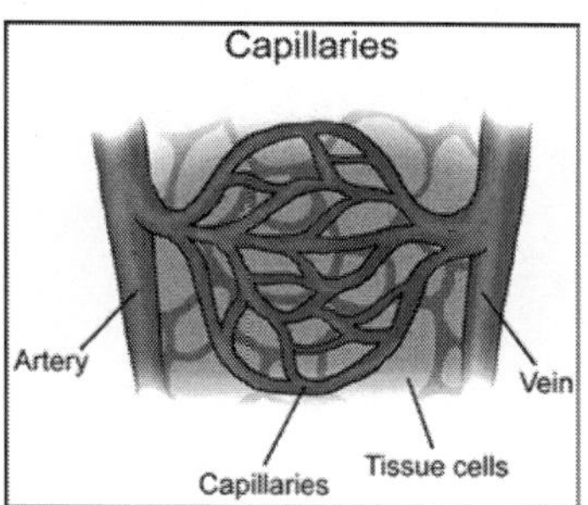

Fig. Interface where Arteries Split into Capillaries, Gases are Exchanged with the Nearby Cells, and then the Capillaries Join together to Form Veins that take the Deoxygenated Blood Back to the Heart.

Capillaries are the tiniest blood vessel, and they are made of epithelial cells. Capillaries are the site of gas and nutrient exchange, and to facilitate diffusion the wall of a capillary is one cell thick. A capillary is approximately 1 mm in length and it is so narrow that red blood cells have to go through single file. Veins carry blood returning to the heart, and since the blood they are carrying is under less pressure, they are built differently than arteries. Veins have thinner walls and larger diameters, which encourage the blood to flow back to the heart. Veins have one-way valves incorporated into them, which prevent blood from flowing backwards away from the heart.

Consists of Blood Cells

In vertebrates, blood consists of cells, cell fragments, and plasma. When blood is separated in a centrifuge (or by letting it sit long enough in a test tube) you can see a yellow fluid at the top and a dark red mass at the bottom. The top fluid is plasma, which contains various nutrients, dissolved gasses, and special immune system proteins. The bottom part contains white blood cells, platelets, and red blood cells.

White blood cells are body's militia. They patrol the body to defend it from pathogens like viruses and bacteria. Platelets are cell fragments that are key ingredients in forming a blood clot. If you cut yourself and start to bleed, the platelets will form a net-like structure at the site of the injury and prevent blood from freely flowing out. Together with other proteins found in the plasma, they will plug the injured site until it can be properly repaired.

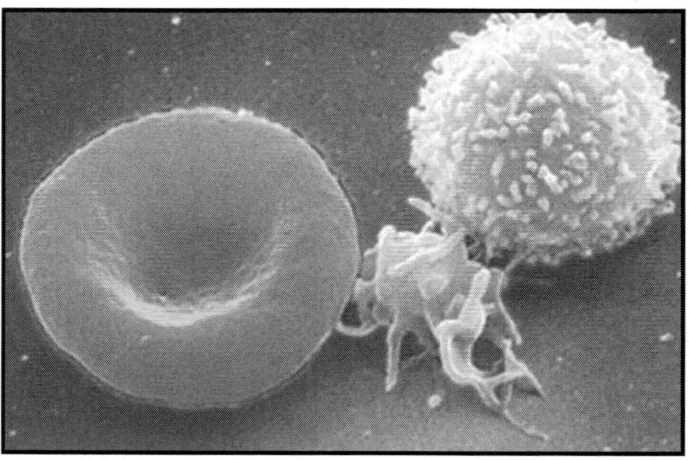

Fig. Three Types of Blood Cells: Red (Left), Platelet (Middle), and White (Right).

Red blood cells are round disc shaped cells whose primary function is carrying oxygen to the cells and carrying carbon dioxide away from them. Red blood cells are filled with hemoglobin, which is the key protein for gas exchange. It is also the reason why blood is red. Hemoglobin is high in iron, which gives the red colour. When a red blood cell is carrying oxygen it is bright red in colour. When it is deoxygenated it takes on more of a purple colour, which is why the veins in your wrist look blue.

Hemolymph in mollusks uses hemocyanin instead of hemoglobin to carry oxygen, and oxygenated hemocyanin turns blue. Unlike hemoglobin, hemocyanin is not part of a red blood cell and is suspended on its own within the hemolymph. Insect hemolymph does not need to carry oxygen since insects have their own system for gas exchange—more on that later.

Pathway of Blood Flow

In a closed circulatory system, the blood can travel in either a single circulation pathway, as seen in fish, or it can travel in a double circulation pathway. The two pathways differ in how many times the heart pumps the blood during a single cycle of circulation.

Fish have a single circulatory system in which the blood is pumped from the heart to the gills and then to various points in the body. For every cycle though the body it goes to the heart one time.

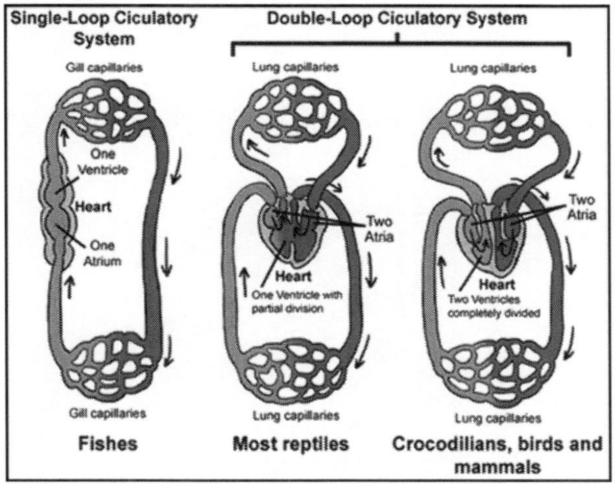

Diagram of the Differences in Animal Circulatory Systems

In fish, the first destination for the blood is the gills. This is the Shell gas station of the fish; it's where the blood loads up on oxygen. Blood flowing through the gills is especially slow because of the tiny narrow capillaries and the massive gas exchange taking place. This loss of momentum is the downside to a single circulation system, because when the blood exits the gills it just barely has enough pressure to make it around the rest of the body.

We've come to the portion of the guide where we compare Michael Phelps to blood cells. Imagine Mr. Phelps pushing off one wall of a pool of a small pool and trying to reach the other side without kicking. He leaves with a lot energy and expects to reach the other side with ease (don't we all?)...BUT, then he travels through a tight tunnel that slows him down. Once free of the tight tunnel, Mike has lost most of his momentum, but he is still able to slowly glide to the other end of the pool. In larger animals it would be like trying that with a giant pool. A single pump is just not enough to send the blood to the lungs, around the body, and back again. Instead, mammals, birds, reptiles, and amphibians have a double circulation pathway. This means that the heart pumps the blood twice per circulation.

The first pump is for pulmonary circulation which specifically sends the blood to the lungs to get oxygenated and then it returns it to the heart. The second pump is for systemic circulation which sends the oxygenated blood out to the rest of the body.

This teamwork provides enough pressure for the blood to get through the lungs and also make it through the rest of the body. The movement through the heart in the double circulation system always starts in the right atrium and the proceeds to the right ventricle. The right ventricle sends the blood to the lungs, which slows it down. However, this blood returns to the heart where it enters the left atrium. The final chamber of the heart that the blood goes to is the left ventricle, which gives it a final push and sends it for systemic circulation out to the far corners of the body.

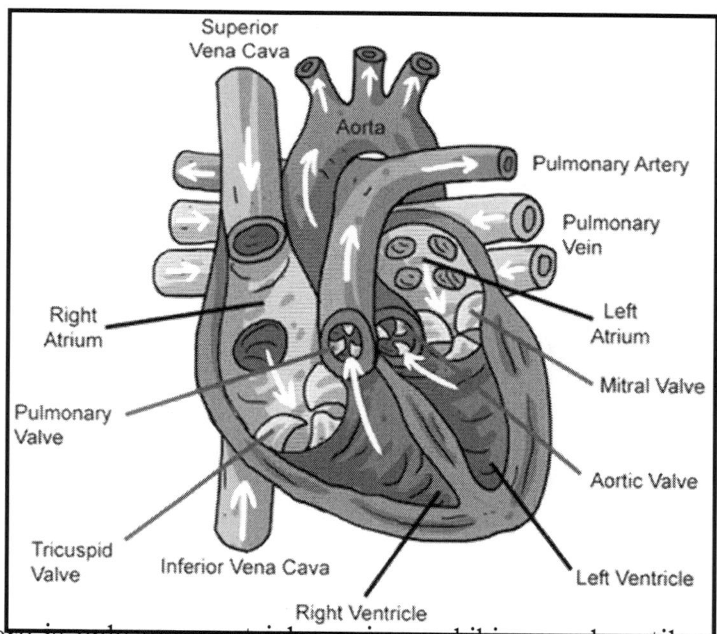

If there is only one ventricle, as in amphibians and reptiles, the same pathway is followed, except that the right and left ventricles are a single cavity. This means that oxygenated and non-oxygenated blood can mix together, and that the blood being sent out towards the body will likely only be partially oxygenated. There is sometimes a small fence-like divider of tissue called a septum, which partially divides the heart and prevents some of the blood from mixing.

Uses of the Circulatory System

The circulatory system is used for delivering oxygen and nutrients to the cells. When we exercise our muscles, we are using oxygen very quickly and as a result our cells need us to speed up the delivery cycle. This is why our breathing and heart rate increases when we exercise.

The circulatory system is also important for removing wastes from the cells. In many animals the waste is delivered to the kidneys and the liver where it is processed for excretion from the animal. The circulatory system is also involved in temperature regulation.

FUNCTIONS OF CIRCULATORY SYSTEM

The functions performed by the circulatory system in all animals are, Transport of nutrients like glucose, amino acids, vitamins, minerals.

Types of Circulatory System

In the closed type of circulatory system, the blood remains inside the blood vessels and does not come out. The blood flows from arteries to veins through small blood vessels called capillaries. The closed type of circulatory system occurs in most of the Annelids, Cephalopods and Vertebrates including man.

Open Circulation System in Cockroach

The heart of the cockroach is elongated, thick, muscular, tubular and 13-chambered. It lies in the pericardial sinus of the haemocoel. Each chamber of the heart receives oxygenated blood from the dorsal sinus through one pair of slit like openings called ostia.

The heart contracts in a postero anterior direction and the blood also flows posteroanteriorly. The alary muscles are responsible for the circulation of blood. The first chamber leads into an aorta, which opens in the head sinuses which are connected to the pericardial sinus through perineural and perivisceral sinuses.

Closed Circulatory System in Humans

The organisms have a thick body wall to prevent the evaporation of water, so exchange of materials between the body cells and the environment by diffusion is not possible.

Blood Vessels

These are hollow, tubular vessels which conduct the blood from the heart to the tissues and from the tissues to the heart.

There are 3 type of blood vessels:

- Arteries,
- Capillaries and
- Veins.

Components of Circulatory System

The WBCs are larger in size than the RBCs but much lesser in number. They are different from the erythrocytes in the following aspects.

Heart—Shape and Position

It is a thick muscular, reddish brown, conical organ present in the mediastinal space of thoracic cavity between 2 pleura enclosing the lungs. Its broader side is called base and it is forward and upward while the pointed side called apex is backward and downward. It is 9 cm broad and 12 cm long and about 300 gms in weight.

Heart—Cardiac Cycle

This phase involves the contraction of the 2 auricles, pushing the blood into the respective ventricles. There is no back flow of blood due to the presence of the bicuspid and the tricuspid valves. The atrial systole takes 0.1 second. This is followed by the atrial diastole when both the auricles relax simultaneously. This is about 0.7 seconds.

Circulation of Blood through the Mammalian Heart

The aorta or the great artery arising from the left ventricle takes the oxygenated blood to the body organs through a number of arteries. From these organs, the deoxygenated blood is collected by the superior and inferior vena cava and brought back to the right auricle. The right auricle pumps the deoxygenated blood into the right ventricle.

Nervous Regulation

The heart is regulated to a large extent through the central nervous system. The heart nerve branches from the vagus nerve from the medulla oblongata and the sympathetic nerve fibres from the spinal cord.

Rate of Heart Beat

Pulse is defined as the wave of distension and recoiling felt in the radial artery due to the contraction of the left ventricle which force about 70-90 ml of blood into the already full aorta.

Causes of Blood Pressure

The pressure existing in the arteries is called arterial blood pressure. The blood pressure is high during systole and is called systolic blood pressure. It is low during diastole and it is called diastolic blood pressure. The difference between these two pressures is called pulse pressure.

Blood Related Disorders

Persistently having a resting blood pressure of more than 120 / 80 mm Hg is called hypertension. In such a condition, the heart has to work harder to pump the required amount of blood to the various organs through narrowed arteries.

ECG - Electrocardiography

ECG is a graphic display or recording of the electrical variations produced by heart muscle during a cardiac cycle. The activity of the heart create an electrical field which is conducted through the surrounding body tissues to the surface of the body.

Components of Human Lymphatic System

The lymphatic system is an accessory circulatory system which transports lymph, a fluid similar to plasma from the intercellular spaces of tissues to the blood. It is a one way route for the movement of interstitial fluid to blood. The lymphatic system can carry proteins and large particle matter from the tissue spaces into the blood.

Immunity and Immune System

Animals encounter many potentially dangerous microbes in air, water and food. So they have involved defence mechanisms against the disease causing germs Immunity is a defence mechanism without which we will fall a prey to all parasitic microorganisms.

Internal Defence

Second line of defence or body's internal defence is carried out by leucocytes, macrophages, inflammatory reactions, fever, interferons and natural killer cells. These operate together to check the damage to the body by the pathogens.

Specific Defence Mechanism

This mechanism provides protection against specific foreign materials and is often called the immune system. This system forms the third line of body's defence against microbes and harmful molecules. The most important characteristic of the immune system is that its cells have an ability to recognise body's own cells and macromolecules, from those which are foreign invaders or non-self.

Cells of the Immune System

Lymphocytes are the main cells of the immune system. They arise from stem cells present in the liver in the foetus and in the bone marrow in the adult.

Humoral Immune System

The plasma membrane of each B cell should be sensitised by contact with a specific antigen for the release of antibodies. The plasma cells do not migrate to the site of infection but act through the lymph. So they form the humoral immune system. The B lymphocytes are short lived and are replaced every few days from the bone marrow.

Cell Mediated Immune System

The cellular immune response is given by T - cells. There are separate T - cells for each type of antigen that invades the body. The life span of the T - cells is 4-5 years or even longer. There are four types of T - cells.

Immunity

Immunity is the ability of an organism to recognise the foreign material or chemicals that enter the body and to mobilise the cells and cell products to quickly remove the foreign material.

Types of Immunity

The type of immunity inherited by the organism from the parents and protects it from birth throughout life is known as innate or inborn immunity.

ANIMAL CIRCULATORY SYSTEMS

An efficient circulatory system has:

- A fluid, *e.g.*, blood, to carry the materials to be transported;
- A system of vessels to distribute the blood;
- A pump to push the blood through the system;
- Exchange organs to carry out exchanges between the blood and external environment, *e.g.*,
  - Lungs and intestine to add materials to the blood;
  - Lungs and kidneys to remove materials from the blood.

- The most crucial demand on the circulatory system is the transport of oxygen and carbon dioxide to and from a gas exchange organ:
  - Lungs or
  - Gills and
  - Tissues.
- All exchanges between blood and cells occur in the capillaries.
- The force of the pump that pushes blood through the arteries is dissipated as the blood flows through capillaries. Although capillaries are tiny, the total cross-sectional area of all the capillaries supplied by a single artery is much greater than that of the artery itself. Like a rapid, narrowly-confined stream spreading out over a flat plain, the force and velocity of flow diminish quickly.

This creates a problem:

- If the pump is used to deliver blood with force to the gas exchange organ, little force remains to distribute the oxygenated blood to the tissues.
- If the pump is used to deliver blood with force to the tissues, little force remains to send the deoxygenated blood to the gas exchange organ.

The Fish Heart

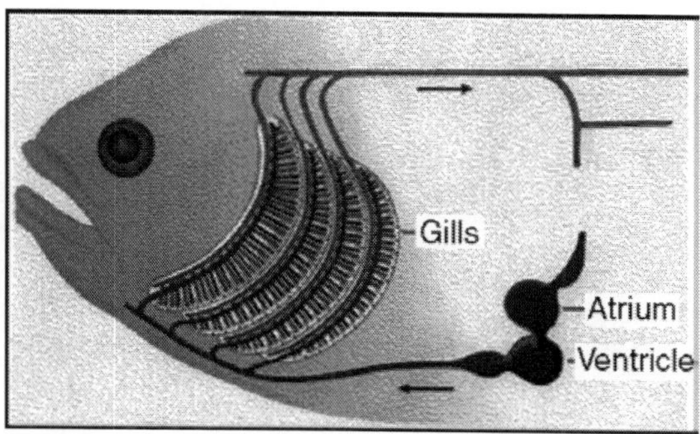

Most fishes have never solved this problem, which is probably why most of them are "cold-blooded".

- Blood collected from throughout the fish's body enters a thin-walled receiving chamber, the atrium.
- As the heart relaxes, the blood passes through a valve into the thick-walled, muscular ventricle.
- Contraction of the ventricle forces the blood into the capillary networks of the gills where gas exchange occurs.
- The blood then passes on to the capillary networks that supply the rest of the body where exchanges with the tissues occur.

- Then the blood returns to the atrium.

While obviously adequate to the fish's needs, this is not a very efficient system. The pressure generated by contraction of the ventricle is almost entirely dissipated when the blood enters the gills.

The Squid Hearts

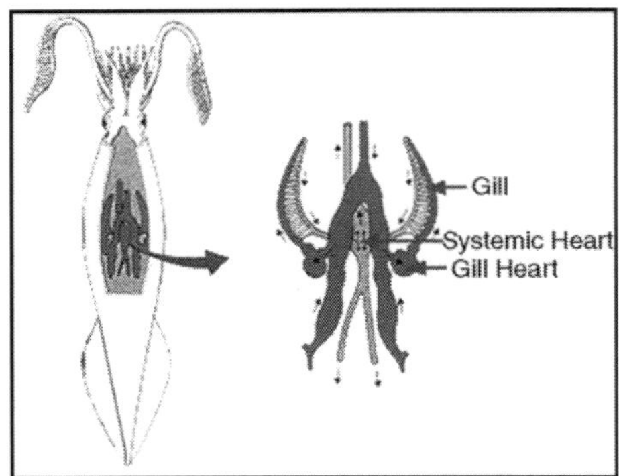

This group of marine invertebrates has solved the problem by having separate pumps:

- Two gill hearts to force blood under pressure to the gills and
- A systemic heart to force blood under pressure to the rest of the body.

Three Chambers: The Frog and Lizard

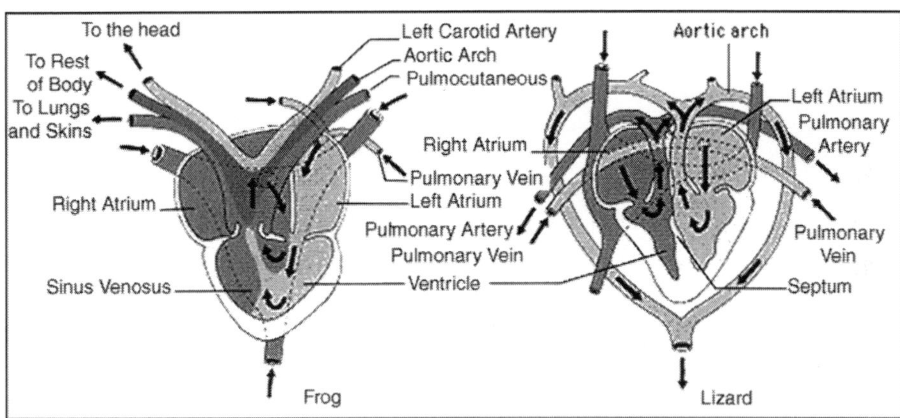

The Frog Heart

The frog heart has 3 chambers: two atria and a single ventricle.

- The atrium receives deoxygenated blood from the blood vessels (veins) that drain the various organs of the body.

- The left atrium receives oxygenated blood from the lungs and skin (which also serves as a gas exchange organ in most amphibians).
- Both atria empty into the single ventricle.
- While this might appear to waste the opportunity to keep oxygenated and deoxygenated bloods separate, the ventricle is divided into narrow chambers that reduce the mixing of the two blood.
- So when the ventricle contracts,
  - Oxygenated blood from the left atrium is sent, relatively pure, into the carotid arteries taking blood to the head (and brain);
  - Deoxygenated blood from the right atrium is sent, relatively pure, to the pulmocutaneous arteries taking blood to the skin and lungs where fresh oxygen can be picked up.
  - Only the blood passing into the aortic arches has been thoroughly mixed, but even so it contains enough oxygen to supply the needs of the rest of the body.
- Note, that in contrast to the fish, both the gas exchange organs and the interior tissues of the body get their blood under full pressure.

The Lizard Heart

- Lizards have a muscular septum which partially divides the ventricle.
- When the ventricle contracts, the opening in the septum closes and the ventricle is momentarily divided into two separate chambers.
- This prevents mixing of the two bloods.
  - The left half of the ventricle pumps oxygenated blood (received from the left atrium) to the body.
  - The right half pumps deoxygenated blood (received from the right atrium) to the lungs.

FOUR CHAMBERS: BIRDS, CROCODILES, AND MAMMALS

The septum is complete in the hearts of birds, crocodiles, and mammals providing two separate circulatory systems:

- Pulmonary for gas exchange with the environment and
- Systemic for gas exchange (and all other exchange needs) of the rest of the body.

The efficiency that results makes possible the high rate of metabolism on which the endothermy ("warm-bloodedness") of birds and mammals depends.

TYPES OF CIRCULATORY SYSTEMS

Living things must be capable of transporting nutrients, wastes and gases to and from cells. Single-celled organisms use their cell surface as a point of exchange with the outside environment.

Multicellular organisms have developed transport and circulatory systems to deliver oxygen and food to cells and remove carbon dioxide and metabolic wastes. Sponges are the simplest animals, yet even they have a transport system.

Seawater is the medium of transport and is propelled in and out of the sponge by ciliary action. Simple animals, such as the hydra and planaria, lack Specialised organs such as hearts and blood vessels, instead using their skin as an exchange point for materials. This, however, limits the size an animal can attain. To become larger, they need Specialised organs and organ systems.

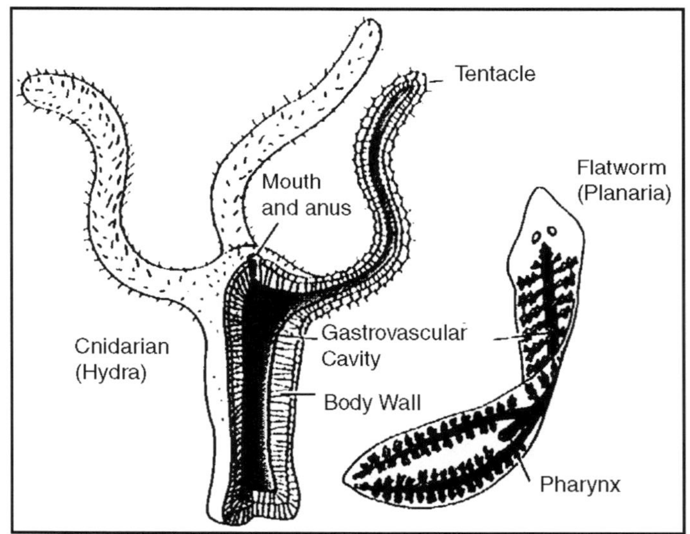

Fig. Structures that Serve Some of the Functions of the Circulatory System in Animals that Lack the System.

Multicellular animals do not have most of their cells in contact with the external environment and so have developed circulatory systems to transport nutrients, oxygen, carbon dioxide and metabolic wastes. Components of the circulatory system include

- *Blood:* A connective tissue of liquid plasma and cells
- *Heart:* A muscular pump to move the blood
- *Blood vessels:* Arteries, capillaries and veins that deliver blood to all tissues

There are several types of circulatory systems. The open circulatory system is common to molluscs and arthropods. Open circulatory systems pump blood into a hemocoel with the blood diffusing back to the circulatory system between cells. Blood is pumped by a heart into the body cavities, where tissues are surrounded by the blood. The resulting blood flow is sluggish.

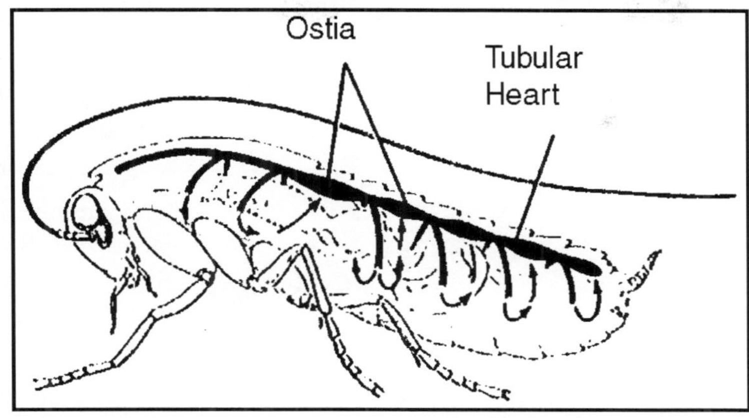

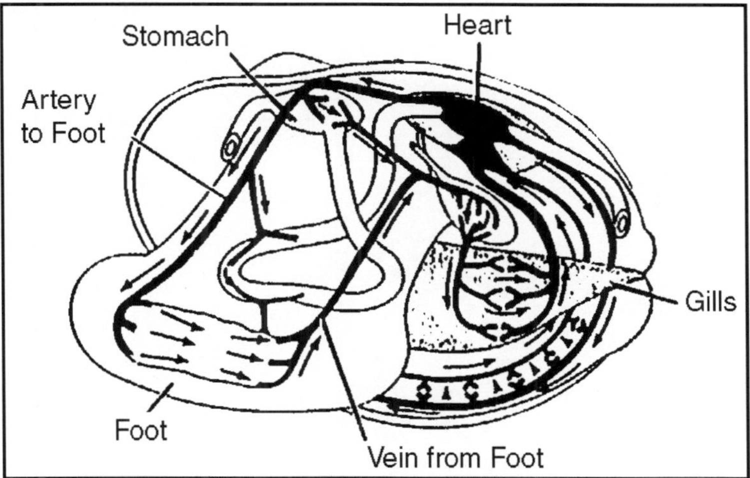

Fig. Circulatory Systems of an Insect (Top) and Mollusc (Middle).

Vertebrates, and a few invertebrates, have a closed circulatory system. Closed circulatory systems (evolved in echinoderms and vertebrates) have the blood closed at all times within vessels of different size and wall thickness. In this type of system, blood is pumped by a heart through vessels, and does not normally fill body cavities.

Blood flow is not sluggish. Hemoglobin causes vertebrate blood to turn red in the presence of oxygen; but more importantly hemoglobin molecules in blood cells transport oxygen.

The human closed circulatory system is sometimes called the cardiovascular system. A secondary circulatory system, the lymphatic circulation, collects fluid and cells and returns them to the cardiovascular system.

VERTEBRATE CARDIOVASCULAR SYSTEM

The vertebrate cardiovascular system includes a heart, which is a muscular pump that contracts to propel blood out to the body through arteries, and a series of blood vessels.

The upper chamber of the heart, the atrium (pl. atria), is where the blood enters the heart. Passing through a valve, blood enters the lower chamber, the ventricle. Contraction of the ventricle forces blood from the heart through an artery. The heart muscle is composed of cardiac muscle cells. Arteries are blood vessels that carry blood away from heart. Arterial walls are able to expand and contract. Arteries have three layers of thick walls. Smooth muscle fibres contract, another layer of connective tissue is quite elastic, allowing the arteries to carry blood under high pressure.

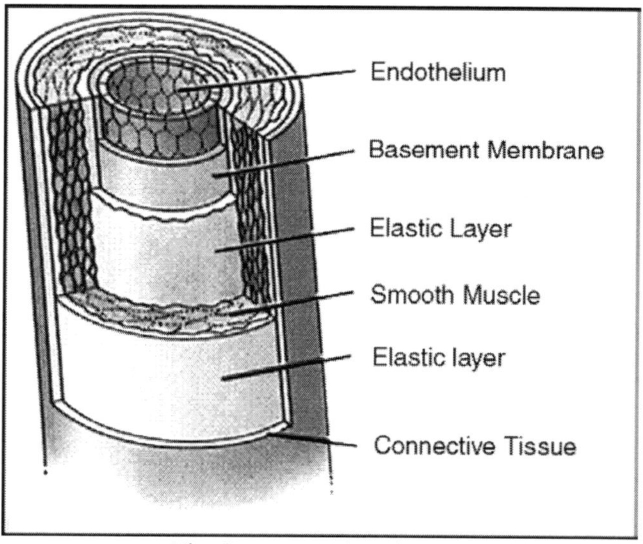

Fig. Structure of an Artery.

The aorta is the main artery leaving the heart. The pulmonary artery is the only artery that carries oxygen-poor blood. The pulmonary artery carries deoxygenated blood to the lungs. In the lungs, gas exchange occurs, carbon dioxide diffuses out, oxygen diffuses in. Arterioles are small arteries that connect larger arteries with capillaries.

Vertebrate Vascular Systems

Humans, birds, and mammals have a four-chambered heart that completely separates oxygen-rich and oxygen-depleted blood. Fish have a two-chambered heart in which a single-loop circulatory pattern takes blood from the heart to the gills and then to the body. Amphibians have a three-chambered heart with two atria and one ventricle. A loop from the heart goes to the pulmonary capillary beds, where gas exchange occurs. Blood then is returned to the heart.

Blood exiting the ventricle is diverted, some to the pulmonary circuit, some to systemic circuit. The disadvantage of the three-chambered heart is the mixing of oxygenated and deoxygenated blood. Some reptiles have partial separation of the ventricle. Other reptiles, plus, all birds and mammals, have a four-chambered heart, with complete separation of both systemic and pulmonary circuits.

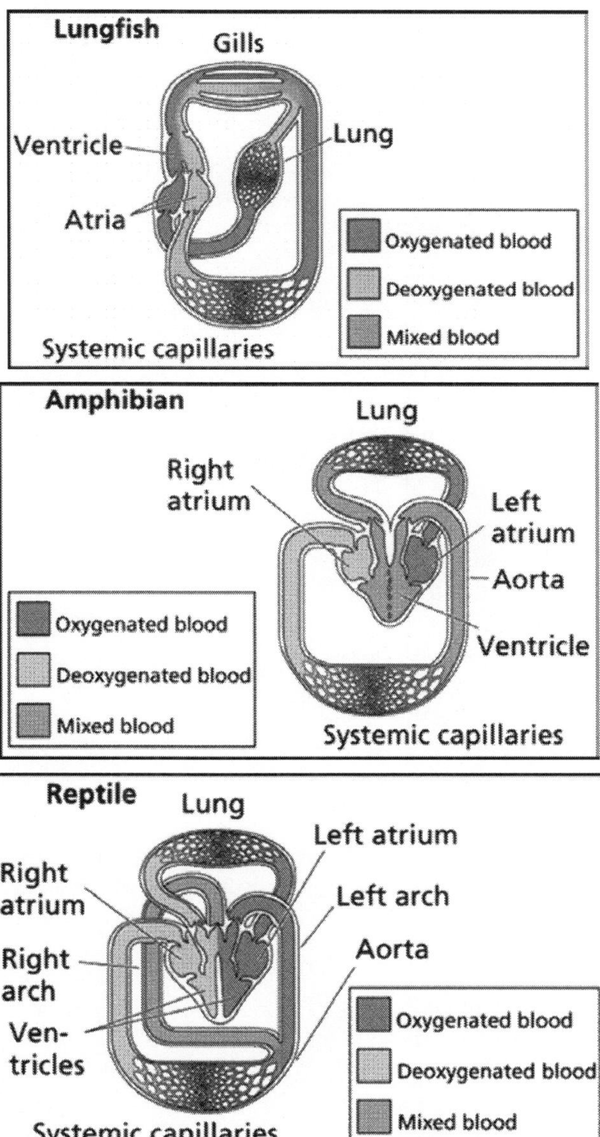

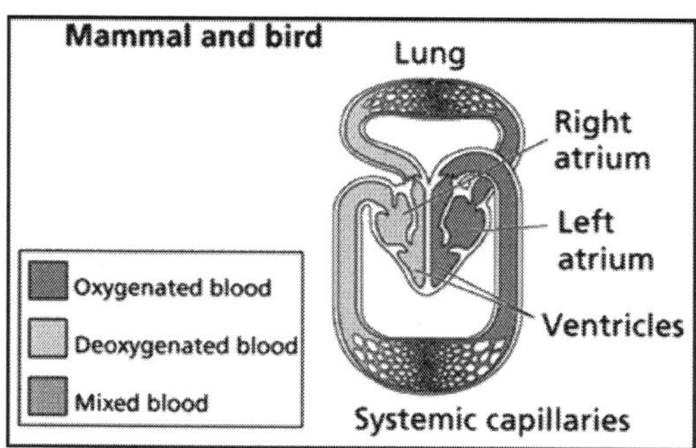

Fig. Circulatory Systems of Several Vertebrates Showing the Progressive Evolution of the Four-chambered Heart and Pulmonary and Systemic Circulatory Circuits.

The Heart

The heart is a muscular structure that contracts in a rhythmic pattern to pump blood. Hearts have a variety of forms: chambered hearts in mollusks and vertebrates, tubular hearts of arthropods, and aortic arches of annelids. Accessory hearts are used by insects to boost or supplement the main heart's actions. Fish, reptiles, and amphibians have lymph hearts that help pump lymph back into veins.

The basic vertebrate heart, such as occurs in fish, has two chambers. An auricle is the chamber of the heart where blood is received from the body. A ventricle pumps the blood it gets through a valve from the auricle out to the gills through an artery.

Amphibians have a three-chambered heart: two atria emptying into a single common ventricle. Some species have a partial separation of the ventricle to reduce the mixing of oxygenated (coming back from the lungs) and deoxygenated blood (coming in from the body). Two sided or two chambered hearts permit pumping at higher pressures and the addition of the pulmonary loop permits blood to go to the lungs at lower pressure yet still go to the systemic loop at higher pressures.

Establishment of the four-chambered heart, along with the pulmonary and systemic circuits, completely separates oxygenated from deoxygenated blood. This allows higher the metabolic rates needed by warm-blooded birds and mammals.

The heart beats or contracts approximately 70 times per minute. The human heart will undergo over 3 billion contraction cycles, during a normal lifetime. Thecardiac cycle consists of two parts: systole (contraction of the heart muscle) and diastole (relaxation of the heart muscle). Atria contract while ventricles relax. The pulse is a wave of contraction transmitted along the arteries.

Valves in the heart open and close during the cardiac cycle. Heart muscle contraction is due to the presence of nodal tissue in two regions of the heart. The SA node (sinoatrial node) initiates heartbeat. The AV node (atrioventricular node) causes ventricles to contract. The AV node is sometimes called the pacemaker since it keeps heartbeat regular. Heartbeat is also controlled by nerve messages originating from the autonomic nervous system.

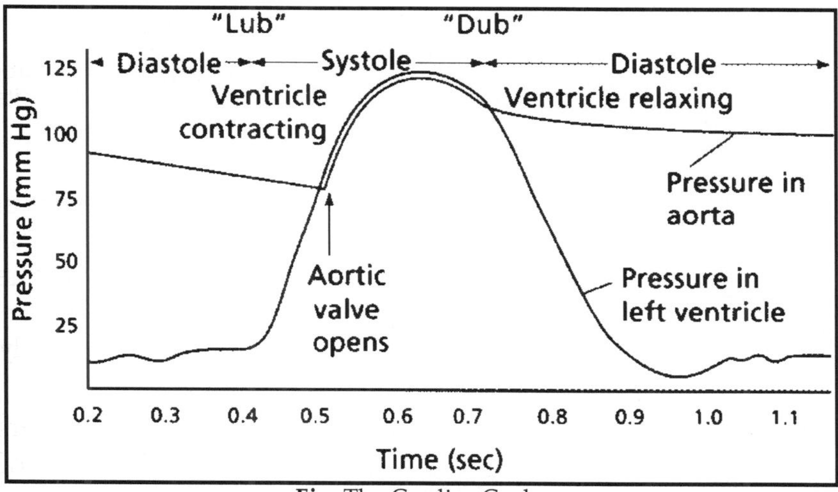

Fig. The Cardiac Cycle.

Blood flows through the heart from veins to atria to ventricles out by arteries. Heart valves limit flow to a single direction. One heartbeat, or cardiac cycle, includes atrial contraction and relaxation, ventricular contraction and relaxation, and a short pause. Normal cardiac cycles (at rest) take 0.8 seconds. Blood from the body flows into the vena cava, which empties into the right atrium. At the same time, oxygenated blood from the lungs flows from the pulmonary vein into the left atrium. The muscles of both atria contract, forcing blood downward through each AV valve into each ventricle.

Diastole is the filling of the ventricles with blood. Ventricular systole opens the SL valves, forcing blood out of the ventricles through the pulmonary artery or aorta. The sound of the heart contracting and the valves opening and closing produces a characteristic "lub-dub" sound. Lub is associated with closure of the AV valves, dub is the closing of the SL valves.

Human heartbeats originate from the sinoatrial node (SA node) near the right atrium. Modified muscle cells contract, sending a signal to other muscle cells in the heart to contract. The signal spreads to the atrioventricular node (AV node). Signals carried from the AV node, slightly delayed, through bundle of His fibres and Purkinjie fibres cause the ventricles to contract simultaneously.

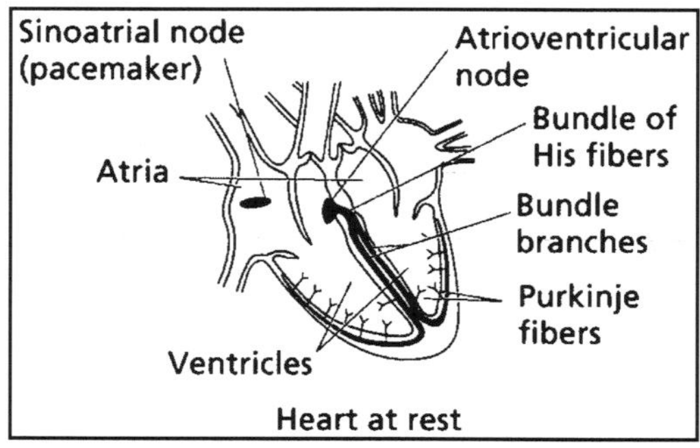

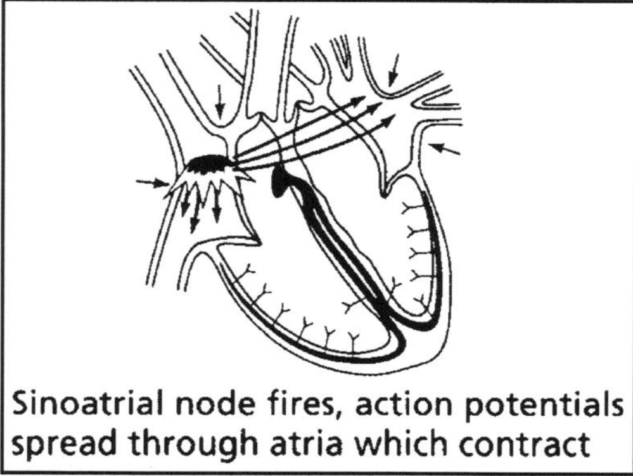

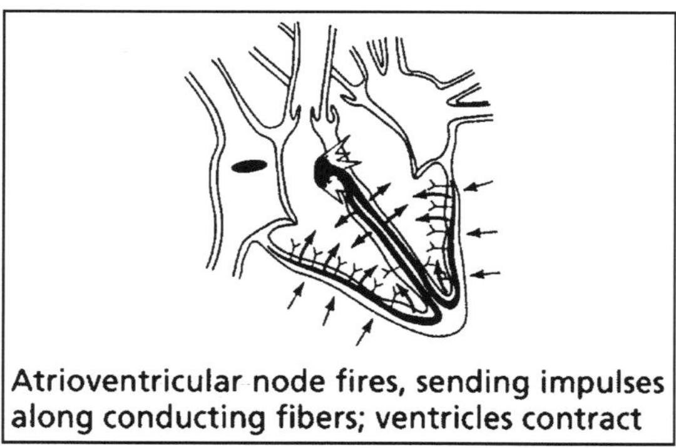

Fig. The Contraction of the Heart and the Action of the Nerve Nodes Located on the Heart.

DOUBLE CIRCULATION

Double circulation is the process by which the cardiovascular systems of many vertebrates such as mammals and birds circulate blood throughout their bodies. In this system, the heart pumps the blood twice to perform its function. The first pumping sends the blood to be circulated though the lungs, and the second pumping circulates the blood throughout the body.

Humans, like most other mammals, have a highly developed circulation system. In humans, the heart is divided into the right and left halves, which are actually two separate pumping and flow systems. The right side of the heart accepts blood that has been circulated throughout the body and pumps it at a relatively low pressure to the lungs. In the lungs, the blood is replenished with oxygen.

Once the blood has been completely circulated through the lungs, it returns to the heart. Arriving at the much larger left side of the heart, the blood is then pumped out to the entire body. As this side must push the blood though a greater amount of tissue, it squeezes much harder than the right, creating a greater amount of pressure. After the blood has been fully circulated throughout the body, it returns to the right side of the heart, where the process is repeated.

Though the blood is considered to be fully circulated upon its return to the heart, each blood cell does not actually visit each part of the body. In double circulation systems, there are many different paths called arteries that the circulating blood may take to reach different parts of the body. There are also many different paths called veins that the blood uses to return to the heart.

The blood is delivered in the quantities needed by each portion of the body and then returned to the heart to be sent for reoxygenating in the lungs. Generally, the more blood needed by a specific part of the body, the more direct the path the blood takes to and from the heart.

Though most highly developed animals have double circulation systems, not all do. Fish, for example, have a single circulation system that lacks the double-sided hearts of mammals. In fish, the blood circulates throughout the body until it reaches the heart. Once it returns there, the blood is pumped again, passes through the fish's gill system, and then circulates on through the body again.

POSTERIOR CIRCULATION STROKE ANIMAL MODELS

The posterior circulation is an understudied brain region that is affected by stroke. When translational research progresses to clinical trials, most trials will enroll very few or completely exclude posterior stroke patients. Though posterior circulation strokes are too uncommon in many population centres to achieve sufficient numbers, other studies try to control for the heterogeneity between the anterior and posterior circulations.

This leads to evidence-based guidelines which may not sufficiently represent some important spectrums of stroke. For these reasons, experimental animal models could be a useful tool to address emerging posterior circulation treatment strategies. In this chapter, we will integrate clinical features with animal models in describing characteristics of posterior circulation strokes, including the neurovascular features, and pathophysiology mechanisms founded from these experimental models.

Hemodynamic Posterior Circulation

The posterior circulation originates from the paired vertebral arteries and a single basilar artery, to supply the inferior thalamus, occipital lobes, midbrain, cerebellum, and brainstem. At the pontomedullary junction, the vertebral arteries fuse to form the basilar artery, which then courses along the ventral aspect of the pons and mesencephalon. From the basilar artery, dorsolateral (circumferential) superficial vessels branch out to the lateral sides and course towards the cerebellum, while deep (paramedian) branches perforate directly into the brainstem, along the ventral aspect.

The basilar artery terminates at the mesencephalic cistern, with perforator branches to parts of the diencephalon, and bifurcation into the paired posterior cerebral arteries (PCAs). The PCAs course laterally to combine with the posterior communicating arteries (PComAs) and then continue to supply parts of the occipital and temporal cortices. The circumferential, paramedian, and other perforator branches are called terminal vascular branches, which lack collateral flow and may potentiate focal ischemia during vertebrobasilar vessel occlusions. The pontine paramedian and the lateral cerebellar circumferential branches are the most common sites of hemorrhage. Several clinical syndromes are described for posterior circulation vascular injuries. Vascular reserve within the basilar circulation includes bidirectional flow through the AICA, PICA, and cerebellar leptomeninges. The leptomeningeal interconnections between cerebellar arteries are similar to the cerebral pial network and can reverse blood flow back through the tributaries of the basilar artery. Outside the posterior circulation, the direction of blood flow can be reversed through hemodynamic connections between PComA (posterior communicating artery), first PCA segment, and carotid circulation.

Increased PComA vessel luminal size is directly proportional to improved patient outcome after basilar artery and first segmental PCA occlusions. Patients with PComAs greater than 1mm in diameter have less ischemic injury during carotid territory occlusions. During basilar artery occlusion, PComAs reverse blood flow through the basilar bifurcation, PCA, and SCA (quadrigeminal plate). However, individual variations in arterial anatomy and the collateral circulation are common (asymmetric or single vertebral arteries, SCA and AICA branching variants, small PComAs) and these can narrow the basilar artery, diminishing vascular reserve, and leading to a greater incidence and severity of stroke.

Transient Posterior Brain Damage

Around ten years ago (vertebrobasilar), transient ischemic attack (TIA) was defined as follows: "a brief episode of neurologic dysfunction caused by focal brain ischemia, with clinical symptoms typically lasting less than one hour, and without evidence of acute infarction". VTIAs are half the duration of carotid territory TIAs and generally perceived by clinicians as having a more benign course.

Consequently, VTIA patients receive less clinical investigation and treatment. However, a systematic review of sixteen-thousand patients found no differences between carotid and vertebrobasilar TIAs, in the rate of stroke, death, or disability. In fact, VTIAs are more likely to convert into full-on strokes during the acute-phase, and a third will have a stroke within 5 years.

Ischemic Posterior Brain Damage

One-quarter of all ischemic strokes are located in the vertebrobasilar (VB) territory. These are usually caused by thrombi/emboli and rarely from vertebral artery dissection of C1-2 vertebral level trauma. Patients with large vessel (basilar artery or intracranial VA) occlusions affecting the brainstem tend to have a worse prognosis while small lacunar occlusions generally do well, so long as cardiorespiratory centres are intact.

Patient outcomes after VB ischemic stroke have been somewhat the subject of debate. The Oxfordshire Community Stroke Project prospectively followed 129 patients and found a 14 per cent mortality and 18 per cent major disability rate, while the New England Medical Centre Posterior Circulation Registry (NEMC-PCR) found a 4 per cent (death) and 18 per cent (disability) rate, with a prospective study of 407 patients. For basilar artery occlusion (BAO), the most severe form of VB ischemic stroke, a systematic analysis of 10 published case series and 344 patients, reported an overall death or dependency rate of 76 per cent, while the NEMC-PCR study with 87 patients reported poor outcomes in 28–58 per cent of patients.

Hemorrhagic Posterior Brain Damage

One-fifth of all intracerebral hemorrhage (ICH) occurs in the cerebellum or brainstem. Brainstem hemorrhages have a 65 per cent mortality rate and around 40 per cent after cerebellar hemorrhage. Prolonged endovascular cerebrovascular damage from uncontrolled hypertension leads to arteriosclerotic and amyloid angiopathic changes, vessel fragility, and rupture at the deep cerebellar vessels or brainstem basilar (paramedian) branches. Less common relations to occurrence are cancer, coagulopathy, or vascular anomalies (arterial-venous malformations, aneurysms, cavernomas, and dural arteriovenous fistulas). For most patients, supportive care is the only treatment rendered, since surgery is only available for one-quarter of hospitalised cerebellar hemorrhage patients, and the brainstem is not surgically accessible. Mechanisms of infratentorial hemorrhage have never been studied and to this end we have developed animal models using collagenase to address this brain hemorrhage subpopulation.

Animal Studies

Experimental models are available to study ischemic posterior circulation stroke. Many animal studies of anterior circulation ischemic stroke have demonstrated impaired autoregulation after ischemic stroke. The extent of which would depend on occlusion duration and extent of reperfusion hyperemia. These mechanisms warrant further study—this can be achieved using available animal models of posterior circulation stroke.

Under experimental conditions, the standardised progressive hypotension in rats showed that autoregulatory kinetics remained intact at the cerebrum, while a progressive loss of autoregulatory efficacy in the cerebellum. As a next step, however, changes in mean arterial blood pressure (MABP) and CO_2 levels (in cats) while measuring blood flow (hydrogen clearance method) in the cerebrum, cerebellum, and spinal cord found greater susceptibility to pressure-dependant ischemia in the cerebrum and spinal cord than cerebellum, which was relatively resistant.

Corroborative studies used transcranial Doppler methods for comparing blood flow in supratentorial and infratentorial brain compartments during increasing intracranial pressures, in the rabbit experimental model. Essentially, the maximum vasomotor activity amplitude of occurred 30 seconds later in the basilar artery, compared with the carotids. Such reports demonstrate that delays are present in the effect of intracranial pressure upon hindbrain microvascular tone.

Using a canine experimental model of permanent occlusion to posterior cerebral artery perforators with the ability to monitor cerebral blood flow (autoregulation) and carbon dioxide reactivity, in response to induced hypotension/hypertension, it was found that cerebral cortex maintained autoregulation and carbon dioxide reactivity, while thalamic autoregulation was maintained in hypotension, but not during episodic hypertension.

On the other hand, the midbrain retained marked impaired autoregulation and carbon dioxide reactivity. Such findings reveal differential brain vulnerability following permanent vascular occlusions. In essence, animal studies indicate that brainstem nuclei decompensated compared to forebrain regions, despite abundant amounts of posterior collateral circulation. The animal model of bilateral carotid ligation using spontaneously hypertensive rats showed impaired autoregulation in the cerebrum. However, the addition of stepwise drops in mean arterial pressures caused impairment of cerebellar autoregulation as well.

Hypothetically, it is possible that vulnerability to hypotension in areas distant from the stroke ictus is modulated by alpha-adrenoceptor (vasoconstrictive) neurons responding to cerebral (transtentorial) hypertension signals. The collateral vascular compensation may be a function of age, since bilateral carotid occlusion causes greater dependence upon basilar flow in adult rats, compared to dependence upon extracerebral midline collaterals in younger experimental animals.

Vascular Responses to Stroke of Animal

Experimental studies reveal that similar cerebrovascular mechanisms are found after ischemic and hemorrhagic stroke. Normally, cerebrovascular autoregulation maintains optimal brain tissue perfusion through arterial constriction/dilation in response to local levels of CO_2 and systemic variations of blood pressures (MABP). Human stroke leads to damaged cerebral autoregulation capacity and greater dependence upon systemic arterial pressure occurring after both carotid and vertebrobasilar-based vascular territories.

This impairment is recognised as an important mechanism of secondary brain injury and edema formation, following human ischemic stroke and intracerebral hemorrhage.

There is a rationale behind the tight hemodynamic and respiratory control in the intensive care units.

Animal studies show that the vertebrobasilar vessels have a greater capacity to mechanically vasodilate and vasoconstrict compared to carotid-based vasculature, suggesting greater dynamic autoregulatory ability. This may be a mechanism enabling the hindbrain to divert blood flow to the carotid system during cerebrovascular strain, since drops in total brain perfusion lead to proportionally greater diminished flow across the basilar compared to the middle-cerebral artery.

When systemic CO_2 and blood pressure changes are superimposed upon permanent posterior cerebral artery occlusion, in dogs, this showed graded autoregulatory decompensation caudally from the supratentorial region to the brainstem, while carotid-based autoregulation was preserved. Experiments in rats show cerebral sparing, while systemic hypotension causes progressive decline in cerebellar autoregulatory kinetics, and carotid autoregulatory kinetics remain intact.

The impairment of cerebellar autoregulation also occurs after bilateral carotid ligation in spontaneously hypertensive rats. Conversely, the combination of hypocapnia with systemic hypotension, in cats, caused greater ischemic susceptibility in cortical brain-regions compared with the cerebellum. Cerebellar autoregulatory kinetics may, therefore, accommodate CO_2 fluctuation more favourably, in the face of hypoperfusion, while drops in arterial pressures, without systemic CO_2 change, would affect the cerebellum more severely.

In most species, the cerebellum and brainstem have an abundance of white matter tracts. Magnetic resonance imaging (MRI) perfusion and diffusion studies in humans have determined white matter to have an infarction threshold of 20mL/100g/minute, while gray matter can sustain flow down to infarctions starting at 12mL/100g/minute. A greater density of white matter tracts in the hindbrain would imply greater vulnerability to ischemic injury. Therefore the viability of brainstem cardiorespiratory centres during periods of severe systemic hypotension, global cerebral ischemia, and cardiac arrest will necessitate further study.

Neural Consequences from Stroke of Animal

Animal models show that ischemic interruption of cerebral blood flow leads to hypoxic and anoxic brain injury, increased neuronal excitability, and cell death. Reperfusion injury further augments this damage through free radial production and mitochondrial dysfunction and similar mechanisms are at play after hemorrhagic stroke also.

Neurons in the CA1 hippocampal region are particularly vulnerable to ischemia; yet, experimentally, these cells are more resistant to damage than several areas of the hindbrain. Electrophysiological studies after hypoxic injury have shown greater neuronal excitability in the hypoglossal (CNXII) and dorsal vagal motor (DVMN) cranial nuclei of the brainstem compared to hippocampal CA1 regions.

Animal models show that anoxia of the hypoglossal nucleus will have both greater initial injury and impaired recovery compared with these temporal lobe neurons. *In vitro* simulation of ischemic reperfusion injury, using cell cultures of oxygen-glucose deprivation followed by reoxygenation (OGD-R), showed greater free-radical injury (lipid peroxidation) and mitochondrial impairment in cerebellar cells compared to cerebral cortical cell culture. Experiments comparing cerebellum with brainstem injury, after vertebral arterial occlusions in gerbils, showed greatest amount of cell death near regions controlling coordination and balance (cerebellar interpositus and lateral vestibular nuclei), while brainstem cardiorespiratory areas remain relatively more intact. The scattered mosaic nature of brainstem nuclei means that this is not simply a redistribution of blood flow and is likely a feature of the neuronal environment, and this deserves further study. Experimental studies reveal significant cerebellar fastigial nuclei (FN) involvement in the regulation of blood pressure and flow. This occurs via integration of autonomic signals from vestibular and cerebellar Purkinje neurons.

FN also modulate the function of adjacent medullary structures and autonomic spinal intermediolateral column neurons. In primates, these nuclei interconnect with vestibular (lateral and inferior), reticular (lateral, paramedian, and gigantocellular), and cervical spinal anterior gray neurons. Animal models demonstrate that electrical stimulation of the FN leads to pressor responses with tachycardia, as mediated by fibres passing through, or very close to, the FN, while chemical activation causes a depressor response, with bradycardia via intrinsic FN neuronal activity.

Taken together, cerebellar fastigial nuclei serve important cardiovascular functions, the manner of which is of significant clinical interest, since cerebellar injury in association with cardiopulmonary consequences is a common occurrence.

Neurons of the area postrema (AP) also contribute to cardiovascular regulation. Biochemically, the cell-surface receptors of circulating molecules: angiotensin II (AT1), and vasopressin (V1), are expressed within this brain region.

Here, the angiotensin II neurohormone can reset the baroreflex to higher blood pressure levels through indirect interactions with the nucleus of the solitary tract and interconnections within the medulla. These nuclei can also modulate the cardiovascular regulatory effects of other neuropeptides—such as vasopressin.

While this homeostatic effector readily binds somatic V2 type receptors, causing peripheral vasoconstriction, V1 receptor binding-interactions within the area postrema will paradoxically enhance baroreflex sensitivity towards activation at lower threshold pressure set-points. All together, these pathways help keep the balance of complex cerebrovascular systems.

Development and Gender

Young children exhibit sex differences in the autoregulatory capacity between anterior and posterior circulations. Female children, ages 4–8 years, have higher flow velocities for both the middle cerebral and basilar arteries, while both sexes exhibit greater flow velocity in the middle cerebral compared to basilar arteries. Later, autoregulatory capacities begin to emerge with females (10–16 years old) having greater capacity in the basilar artery than males, but males having the advantage of greater MCA autoregulatory index. Up through adolescence, however, females continue to have higher flow velocities (compared to males) for both the middle cerebral and basilar arteries. This may indicate a gender-specific ability to handle an occlusive thrombus in the hindbrain. Further studies are needed to understand these gender differences.

Bibliography

A.C. Shuttleworth and R.H. Smythe : *Clinical Veterinary Surgery*, Greenworld Publication, Delhi, 2000.

Ajay Kumar Upadhyaya : *Text Book of Preventive Veterinary Medicine*, International Book, Delhi, 2005.

Ashis Kumar Ghosh : *Ethnomedicine for Human and Veterinary Development*, Daya Publication, Delhi, 2009.

Debasis Jana and Nilotpal Ghosh : *Essentials of Veterinary Practice*, Daya Publication, Delhi, 2011.

Dinesh Arora : *Biotech's Dictionary of Veterinary*, Biotech, Delhi, 2004.

F W Nicholas : *Introduction to Veterinary Genetics*, Blackwell Science, 2004.

Gary R. Mullen and Lance A. Durden : *Medical and Veterinary Entomology*, Academic Press an imprint of Elsevier, Delhi, 2013.

Geo F. Boddie : *Diagnostic Methods in Veterinary Medicine*, Greenworld Publication, Delhi, 2005.

Geoffrey Lapage : *Monnig's Veterinary Helminthology and Entomology*, Greenworld Publication, Delhi,l 2000.

Gwilym O. Davies : *Gaiger and Davies' Veterinary Pathology and Bacteriology*, Greenworld, 2000.

Gyanendra Pandey : *Pasvayurveda : Study on Fauna and Veterinary Medicine in Ayurveda*, Sri Satguru Publication, Delhi, 2010.

Har Pal Singh and S.N. Maurya : *Instant Veterinary Drug Index: Biologics and Pharmaceutics*, International Book Distributing Co, 1998.

Indranil Samanta : *Veterinary Bacteriology*, New India Publishing Agency, 2013.

J L Vegad : *A Textbook of Veterinary General Pathology*, International, Delhi, 2012.

J.L. Vegad and A.K. Katiyar : *A Textbook of Veterinary Special Pathology*, IBDCO Publication, Delhi, 2005.

John T. Abrams : *Linton's Animal Nutrition and Veterinary Dietetics*, Greenworld, Delhi, 2000.

M. Mondal and Shailendra Singh: *Question Bank of Veterinary Pathology for Competitive Examinations*, Satish Serial Publishing House, 2015.

M.K. Shukla : *Applied Veterinary Andrology and Frozen Semen Technology*, New India Publication, Delhi, 2011.

M.V. Thrusfield and E.A.M. Graat : *Application of Quantitative Methods in Veterinary Epidemiology*, International Publication, 2007.

Mahesh Kumar and R.D. Sharma : *Textbook of Clinical Veterinary Medicine*, Indian Council of Agri Res, 2009.

Neelesh Sharma and S.R. Upadhyay : *Clinical Veterinary Medicine - II*, New India Publication, Delhi, 2009.

P. Kinjavdekar, H.P. Aithal and A.M. Pawde : *Anaesthesia and Analgesia for Veterinary Graduates*, Satish Serial Publishing, Delhi, 2004.

P. Srinivasan : *Veterinary Anatomy Of the Ox*, Bio-Green Books, Delhi, 2012.

Pawan Kumar Yadav and Mamta Kumari : *Veterinary and Human Health*, Ravi Kant Verma, Ved Prakash Saini, Agrotech Publishing Academy, 2011.

R.S. Chauhan : *Text Book of Veterinary Pathology*, IBDC Publication, Delhi, 2010.

R.S. Chauhan: *Text Book of Veterinary Pathology : Quick Review and Self Assessment*, IBDC Publication, Delhi, 2010.

S.K. Gupta : *Medical, Veterinary and Public Health Important Mites and Ticks*, Nature Books India, 2010.

Vivek M. Patil : *Information Technology in Veterinary Science*, New India Publication, Delhi, 2009.

Index